Praise from the M̶... W9-AZF-579

"Full of down-to-earth advice and support for people whose parents might not have been physically or sexually abusive, but nonetheless limited their children's lives with persistent, unhealthy control."
—*Publishers Weekly*

"Explains how to make sure that your growing pains with your parents don't develop into grown pains."   —*New Woman*

"Doesn't give a cookie-cutter, one-size-fits-all recovery plan, but rather suggests several 'paths to healing' and exercises to help you 'emotionally leave home.' This is self-help at its best."
—Amazon.com

"Accessible, clear, unambiguous. . . . Can help adults heal the wounds of the past, or at least reach a higher understanding."
—*Florida TODAY*

"Neuharth writes and speaks from experience."
—*Columbus Dispatch*

"An often-chilling insight into just how destructive a parent can be."
—*Baltimore Sun*

"Invites readers to discover where they fall on the spectrum, and encourages them to offset the negative effects of such parenting through self-examination, self-responsibility and independence."
—*Marin Independent Journal*,
Marin County, CA

"Explains how overcontrolling parents discourage dissent, stifle strong emotions, criticize more than praise, mete out harsh discipline, make children earn their love, and insist that they're always right."
—*Democrat and Chronicle*, Rochester, NY

"A fresh, compelling, and probing analysis of a lifelong struggle that too few of us confront. Most importantly, the book leads us to an epiphany: more than endure, we can prevail."   —Walter Anderson, author of *The Confidence Course* and executive editor, *Parade*

"It's a terrific book. When the book arrived at the station, I just inhaled it. I loved it."   —Ann Cody, host, *Common Ground*,
WZLX, Boston

"If you wonder whether you were raised by controlling parents, I urge you to get this book. It's a wonderful book."
—Marcia Kimpton, host, *ParenTalk*,
KVON, Napa, California

Praise from Readers

"Your book has really helped me get rid of the feelings of self-hatred and blame I had."   —A.A., Kansas

"Validation! . . . After years of slow, determined self-discovery, finally a book that identifies exactly the condition I suffered in my family."
          —B.G., Illinois

"I found my life on every page."
          —G.H., Colorado

"You've given me enough to work with that I can almost see my way through an upcoming visit to my mother."
          —A.G.

"Your book has given me different insight into my parents' behavior and how I might more effectively deal with it. . . . I have realized that I don't need to apologize to my mother for my career and life opportunities, better relationships with others in my family, and a good life."
          —J.H., Washington, D.C.

"Until your book I didn't feel as though anyone ever understood the depth and intensity of my feelings about being with my parents. . . . I feel validated."   —M.O.

"I was particularly gratified to read the section on forgiveness not being necessary to the healing process—something I'd felt instinctively to be true but felt guilty about."   —M.H., Arizona

"What an enlightening and liberating book! It's the first self-help book I have read where I truly recognized myself immediately. . . . This book gave me an excellent perspective of how controlling behaviors are passed down from generation to generation."
          —D.W.

"You provide hope and you make the reader believe that things can change as you guide the reader to work toward the healing process."
          —F.K., New York

"I turn fifty years old this year and I always felt something was wrong with me. . . . When I opened your book, every word seemed to fit, and my feelings finally made sense. I just had to tell you how much it has meant to me to finally find someone who understands—I'm not alone and I'm not crazy."   —M.M.

IF YOU HAD CONTROLLING PARENTS

How to Make Peace with Your Past and Take Your Place in the World

Dan Neuharth, Ph.D.

HARPER

NEW YORK • LONDON • TORONTO • SYDNEY

A hardcover edition of this book was published in 1998 by Cliff Street Books, an imprint of HarperCollins Publishers.

HarperCollins books may be purchased for educational, business, or sales promotional use. For information please write: Special Markets Department, HarperCollins Publishers Inc., 10 East 53rd Street, New York, NY 10022.

First Cliff Street Books/HarperPerennial edition published 1999.

Reprinted in Quill 2002.

Designed by Nancy Singer Olaguera

The Library of Congress has catalogued the hardcover edition as follows:

Neuharth, Dan.
     If you had controlling parents : how to make peace with your past and take your place in the world / Dan Neuharth.
       p. cm
     ISBN 0-06-019191-0
     Includes bibliographical references.
     1. Adult children of dysfunctional families. 2. Parental overprotection. 3. Manipulative behavior. 4. Control (Psychology) 5. Autonomy (Psychology) 6. Parent and adult child. I. Title
     RC455.4.F3N48 1998
     616.89—dc21   98-8650

ISBN 0-06-092932-4 (pbk.)

  10 11 12 RRD 30 29 28 27 26 25

To the children, past, present, and future,
who lack a voice in their upbringing.
May they find their voices
through the stories and insights shared here

and

To Marly, with love and gratitude
for her artistry, brilliance, beauty, and soul

CONTENTS

Acknowledgments

My profound thanks and appreciation to:

Patti Breitman, superagent, for believing in this book so much and working so brilliantly to sell it.

The incredible staff at Cliff Street/HarperCollins and particularly: Diane Reverand, publisher, for her vision, wisdom, and enthusiasm; Julia Serebrinsky, editor, for her grace and attention to detail; and Pamela Pfeifer, public-relations genius.

The courageous pioneers who volunteered to be interviewed for this book and, in so doing, made new meaning of their own difficult childhoods by helping countless others.

My psychotherapy clients, who teach me more each day.

My steadfast men's group: Scott Cameron, Ph.D.; David Frankel, Ph.D.; Scott Lines, Ph.D.; Mike Shuell, Ph.D.; Alan Vitolo, Ph.D.; and Robert Wynne, Ph.D.

My original family: with love to mother Loretta, father Al, and sister Jan.

My brother-in-law Joseph Keusch, niece Danielle, and nephew Alec, all of whom bring smiles to the world.

Sandy the wonderdog, for offering unconditional love and unlimited play.

My mentors and teachers: Walter Anderson; Robin Acker, M.A., MFCC; Adria Blum, Ph.D.; Bernie Carter, M.A., MFCC; Janeece Dagen, M.A., MFCC; Sandy Graber, M.D.; Roberto Gurza, M.A.; Jerry Schwartz, Ph.D.; and Lucy Scott, Ph.D.

Shannon Tullius and the staff, presenters, and volunteers of the Maui Writer's Conference.

Manuscript readers and supporters: Joan Cox, Lori Hurwitz; and especially Brooke Passano, M.A., MFCC.

And, finally, my partner, Marly Perkins, Ph.D., for her unwavering support, countless hours of reading drafts, and always on-target suggestions. Marly, this book would not exist had you not contributed all your gifts.

Some readers may find that coming to grips with a difficult childhood can spark upsetting feelings. Far from being uncommon, this frequently happens in psychotherapy, which is designed to explore troubling feelings in a safe setting. This book is not intended to be a substitute for formal psychotherapy, though many readers may find it a useful adjunct to treatment. I urge any reader who experiences abnormal depression or anxiety to consult a licensed psychotherapist.

TO THE READER

When I use the word "parents" in this book, I'm talking about the adults who exerted the most significant control over your childhood—birth parents, guardians, grandparents, stepparents, aunts or uncles. I tend to use "parents," plural, for grammatical simplicity even though only one parent or adult figure in your family may have been the controller.

This book includes case studies drawn from comprehensive interviews with a diverse group of forty women and men, ages twenty-three to fifty-eight, who grew up controlled. Collectively, these forty adults have more than six million hours' experience growing up in controlling environments. (Detailed information about the participants and interviews can be found in Notes on Research on pp. 239–240.) Although each person's story was unique, similarities in how they were controlled transcended age, ethnicity, socioeconomic status, gender, sexual orientation, and family history—and strongly mirrored the patterns I've encountered with private clients from controlling families.

I promised confidentiality to all participants so they could talk openly and honestly. Therefore, all names are pseudonyms. I have also slightly altered age, profession, or other details that might identify those interviewed. In some cases, to honor confidentiality I have combined details of more than one person's experience into a composite character. Where needed, I have altered quotes for grammar or clarity. Other than these alterations, every story, incident, and observation you will read was as told to me. None of it is fiction, even though at times the enormity of control may seem unbelievable. I hope the stories and insights will touch you, teach you, and help you to heal, as they have me.

I've learned a great deal about control and healing from my clients and those I interviewed, but there is much I have still to master. I have yet to work out all the mixed feelings and control-related problems between myself and my own parents. You, not I, are the expert on your life, needs, and upbringing. I urge you to take this book at your own pace and on your own terms. I want *you* to feel in control as you read. You don't have to agree with the entire book to find some parts pertinent. And keep in mind that while controlling parents often view the world in all-or-nothing terms, few situations in life are absolute. My goal in writing this book is to remind you that you are not alone, that you can make sense of your childhood, and that you can heal yourself.

This is a book of discovery and resolution. I invite you to discover what may lie underneath some of your most stubborn and troubling habits, patterns, or problems. I urge you to join me on a path toward resolving anything unfinished with your parents, whether they are living or dead. I ask you to fashion a more clear and full view of your upbringing so that you can make peace with your past.

DID YOU GROW UP WITH UNHEALTHY CONTROL?

Animals kill their young if they don't want to care for them, but they don't torture them for years.

—ALICE MILLER

If your parents controlled you in unhealthy ways, they may have unwittingly planted land mines in your psyche. As a result, you may tiptoe through life expecting buried danger, not treasure, in your path. You may wait . . . and wait . . . for permission to love, succeed, and feel content. Permission you're not sure how to get. Permission you may have difficulty granting yourself.

Well, you are not alone. An estimated one in thirteen adults in the United States has grown up with unhealthy control. That's more than 15 million people. (See Notes on Research on pp. 239–240.)

Unhealthy control has lasting costs. Such an upbringing can put you at risk for depression, anxiety, poor self-image, addictions, self-defeating behaviors, and stress-related health problems. Lacking a protective sense of self, you may live with too little freedom, too little meaning, and, most of all, far too little self-love. Growing up controlled means inheriting habits and beliefs that complicate relationships, decision making, spirituality, and emotional development. As one thirty-seven-year-old teacher raised in a white-knuckle household said, "I feel like I'm missing a couple of big chunks on how to be a person."

An unexamined upbringing may lead us unwittingly to replay old patterns with our mates so that our mates come to remind us of our parents. We may misread friends, neighbors, or coworkers who remind us of our parents. We may inadvertently use our children as vehicles to work out unfinished business with our parents. We may unintentionally inflict suffering on ourselves and those around us as we act out old, controlling ways.

After we're grown, our controlling parents may still treat us as children. More frustratingly, we may feel as helpless as children when we're around our parents. We may struggle to get closer to—or find greater distance from—a controlling parent. We may even come to understand their motivation for controlling us, yet be at a loss about reconciling that knowledge with our lingering hurt, disappointment, or anger.

If you have problems or habits that stubbornly resist change, these may be, in fact, symptoms of unresolved issues with your parents or upbringing. For example, we may grow bored with our jobs or relationships when what we may really need is to cut the apron strings with a parent; we may push ourselves mercilessly to do more when what we really need is to slow down and heal old wounds; or we may overeat when what we may really need is to attend to frustrations inherited from childhood. By looking deeper, we can solve these problems at the source so that they don't merely crop up a few months later in a different form.

This book can help you or someone you love to recognize and disarm the emotional land mines that linger from unhealthy family control. I'm here to tell you that many adults who grew up controlled have worked successfully to create happier adulthoods. You'll meet some of them shortly and may find them not all that different from you.

Look at your personality like a puzzle. This book can help you figure out how much of the puzzle was assembled for you by your parents and how many pieces were forced together, whether they fit or not.

How Do You Know?

How do you know if you grew up controlled? Many adults raised with unhealthy control have only a vague sense of it. Others remember excess control but can't explain how it worked. Without something tangible to point to, many who grow up in controlling families come to believe unhealthy control to be normal.

Said a forty-six-year-old designer, "I don't know how to explain it, but my mother had this powerful presence and control. To this day I don't understand how she held so much influence over me or how I took on so many of her values despite my best efforts not to."

Overcontrol takes many forms. The most obvious is authoritarianism, but unhealthy control also occurs in a wide variety of families that are anything but strict. The common factor is this:

Controlling families are organized to please, protect, and serve one or both parents, not to foster optimal growth or self-expression among family members.

This book is for you if you or someone you care about came from a family that could be described as one or more of the following:

- Perfectionistic
- Overprotective
- Dictatorial
- Confusing
- Strict
- Belittling
- Authoritarian
- Manipulative
- Harsh
- Smothering
- Reserved
- Overbearing
- Unyielding
- Tense
- Irritable
- Stifling
- Unemotional
- Pushy

Overcontrol can just as easily exist in a "model" family as in a family having a climate of deception and chaos. Too much control thrives when family members cling to a myth that everything is perfect when it's not. Excessive control can exist when a parent demands too much adulation or insists on iron-clad dos and don'ts. The parent who is too aloof exerts control through deprivation. The parent who is an emotional loose cannon dominates through unpredictability. Overcontrol is fostered by parents who emotionally smother other family members, bully with verbal abuse or physical or sexual violence, or who are too self-absorbed to see their children's needs.

This test will help you measure the prevalence of control in your childhood and identify whether you may be facing adult-life problems because of it. Check all that apply:

GROWING UP, *did you often feel . . .*

○ Forbidden to question or disagree with a parent?

○ Pressured by excessive expectations or unattainable standards?

○ Tense when one or both of your parents were around?

○ Confused by parental mixed messages or unclear rules?

○ Criticized more than you were encouraged or praised?

○ Afraid to express anger, fear, or sadness around a parent?

○ Intimidated or belittled by a parent?

○ Manipulated into doing things you didn't want to?

○ Sad, anxious, hurt, deprived, or angry?

○ That physical and emotional affection were scarce in your family?

○ That pleasing your parents was rewarded more than being yourself?

_____of 11 checked

In RETROSPECT, *did either or both of your parents often . . .*

○ Try to dictate your thoughts, speech, or morals?

○ Overscrutinize your eating, sleep, dress, or personal grooming habits?

○ Interfere with your choices of school, career, friends, or lovers?

○ Violate your privacy?

○ Threaten to disown you for opposing their wishes?

○ Withdraw love or affection when you displeased them?

○ Use words like "lazy," "stupid," "ugly," "selfish," or "crazy" to describe you?

○ Physically or sexually abuse you and/or allow others to do so?

○ Need to be the center of attention or try to dominate most situations?

○ View the world in right-or-wrong, black-and-white terms?

○ Treat emotions as things to be changed, avoided, or ignored?

○ Seem perfectionistic, stoic, or driven?

○ Seem unwilling to admit they were wrong?

○ Seem obsessed with cleanliness, order, details, rules, or schedules?

○ Seem hypersensitive to criticism?

○ Seem unaware of the pain they caused you and others?

____of 16 checked

Did either of your parents . . .

○ Experience major trauma in their childhood?

○ Have a family history of physical or sexual abuse, mental illness, or substance abuse?

○ Feel overcontrolled by their parents?

____of 3 checked

As an ADULT, have you often felt . . .

○ Perfectionistic, driven, or rarely satisfied?

○ Like you are under scrutiny even when no one else is around?

○ Intimidated or easily angered around controlling people?

○ Terrified of being overly dependent in relationships?

○ Strong reservations about having children because of how you were raised?

○ Melancholy, empty, or deprived?

○ Like few people know the real you?

○ Afraid of strong feelings or losing control?

○ That you missed out on large parts of normal childhood experiences?

○ Extrasensitive to criticism?

○ Confused about what your feelings are or should be?

○ Overly judgmental of others?

_____of 12 checked

In your ADULT LIFE, have you often . . .

○ Worried or ruminated over confrontations with others?

○ Found it hard to make decisions?

○ Lost yourself in relationships by putting another's needs first?

○ Had trouble finding a spiritual belief that feels right?

○ Found it difficult to relax, laugh, or be spontaneous?

○ Had difficulty with sex, touch, or intimacy?

○ Had trouble accepting compliments?

○ Had an eating disorder or addictive behavior?

○ Suffered from stress-related illnesses, "burnout," or chronic pain?

○ Undermined yourself in work or relationships?

○ Assumed others have the confidence you lack?

○ Tested the love of those close to you?

○ Been abusive, controlling, or disrespectful to friends or a mate?

○ Expected that others will try to hurt or take advantage of you?

_____of 14 checked

As an ADULT, do you often feel . . .

○ That it has taken a long time to emotionally separate from one or both of your parents?

○ That you visit or talk to a parent more out of obligation than choice?

○ That one or both of your parents don't know you as you really are?

○ That one or both of your parents romanticize your childhood to downplay problems?

○ That you cannot fully please your parents?

○ That your parents just don't get it about their impact on you?

○ Tense when you think about a parent coming to visit?

○ Horrified when you notice yourself acting like one of your parents?

○ A desire to temporarily reduce or sever contact with a parent?

_____of 9 checked

Total Questions: 65

Total Number Checked: ____

If you answered positively to twenty-two or more questions (more than one third the total), you most likely came from a controlling family. Even people from relatively healthy families are going to have some yeses. The difference is that in controlling families, the above tendencies are present more often, to a greater degree, and with greater emotional costs.

However, if you did answer yes to many of the questions, it doesn't mean that you're "damaged goods." It simply means that you faced—and survived—a difficult set of early circumstances that may still affect you. Recognizing this, of course, is the first big step toward healing.

Placing Responsibility

Controlled children rarely have the option of acknowledging, "Something is wrong here. I don't like the way this feels." Because they're trained not to recognize their feelings, controlled children may have only a vague sense of constriction or emotional numbness.

If your parents exerted unhealthy control, something *was* wrong in your family. Healing from such an upbringing often requires that you peek behind the curtain of familial loyalty to examine family rules and beliefs.

Psychoanalyst Alice Miller has written that healing from a painful childhood begins with allowing yourself to express all the feelings and opinions that arose from years of abuse and control; in effect, speaking out after so many years of not being able to.

In so doing, it's important to place responsibility where it truly belongs by acknowledging that:

1. You aren't responsible for what your parents did to you, they are.

2. You are responsible for what you do with your life now, your parents aren't.

Exploring a pattern of control that was handed down for generations in your family isn't passing the buck; it's the first step in stopping the buck. By seeing unhealthy family patterns, you can avoid passing them on—a choice your parents may have been unable or unwilling to make.

This exploration is not designed to blame or bash parents. Being a parent is tough. There is no harder or more important job. Parenting is immensely demanding physically, emotionally, financially, and mentally. No parent gets training in being a parent until she or he becomes one. There are no perfect parents. All parents make mistakes, sometimes big mistakes, and still many children grow up relatively happy, well-adjusted, and able to meet life's challenges.

I do not advocate excessively "permissive" parenting. Appropriate control and limit setting are crucial to child raising. Children test parental control with petulance, sarcasm, deception, and a host of other techniques, some conscious, most instinctive. The lack of adequate limits in permissive households can cause problems no less troubling than the harsh limits in authoritarian families. Yet this book isn't about appropriate control and limit setting, it is about households with *unhealthy control*—too much or the wrong kinds of control for too long.

For most of history, governments have been organized on a patriarchal, authoritarian model. Only recently has democracy, functioning on the consent of the governed, offered an alternative to patriarchal authoritarianism. The first year in which a majority of nations had democratic governments was 1992.

Similarly, most families historically have been based on patriarchal authoritarianism. Of course, a family is not a democracy; children are not yet adults and cannot govern. But I believe both children and parents thrive in "democratic families"—in which both children and adults have the right to speak, think, feel, and trust, free from unhealthy control. As democratic governments become the norm worldwide, how can we expect our children to grow up and live in democracies when they have known only unhealthy control, not democratic ideals?

Control and trust are diametrically opposed and inextricably linked. We control to the extent that we mistrust the world. When we trust the world, we can feel safe enough to let go of much of our need to control. Controlling parents, by and large, do not trust. Parental overcontrol is nearly always a generations-old cycle, in place well before you came along. Most controlling parents, in fact, were themselves tremendously misused as children or were traumatized by family deaths, crises, or abuse. If they never got help for their hurts, they may feel alone in an untrustworthy world, and be desperately trying to control life rather than risk being savaged again.

I feel sad for such controlling parents' deep hurt. Yet parents who ignore or hide their wounds may spend their lives running from the ghosts of the past. In the process, their children pay a tremendous price.

Unlike your parents, you have a choice. You can heal your wounds rather than ignore or hide from them. You can transcend the cycle of control rather than perpetuate it.

The Human Face of Unhealthy Control

As part of researching this book I conducted comprehensive interviews with forty adults ages twenty-three to fifty-eight who grew up controlled. Their experiences illustrate many of the points in this book, weaving a rich tapestry of sadness and hurt, wisdom and hope. You may discover that you have commonalities with many of these people's early experiences as well as with the problems they inherited from unhealthy parental control. Participants ranged from:

- An Arkansas preacher's daughter to the California daughter of a Holocaust survivor

- The daughter of second-generation working-class Italian immigrants to the adopted son of wealthy New England socialites

- The son of Middle Americans whose ancestors fought in the American Revolution to the daughter of Chinese immigrants who barely escaped with their lives in the 1949 Communist revolution

- The oldest daughter of seven children from an Irish-Catholic family to the only daughter of an African-American single mother

- The gay son of a military officer father and fundamentalist Christian mother to the son of a Latin American father and a sadistic, abusive mother

- The daughter whose mother barely survived her childhood in a World War II Japanese concentration camp to the daughter of a schizophrenic mother who could barely negotiate daily life.

Despite their different backgrounds, these people showed striking similarities in how they were controlled and how it affected them. You may discover emotional kinships with some of these people—and you may find that reading about their efforts to heal their difficult childhoods will validate the work you are doing to heal.

There Is Much You Can Do

If your parents were controlling, you saw control modeled as a strategy for living—but it's not the only one. The more aware you are of how your parents controlled and of the fallout of their early control in your present life, the more informed the choices you're likely to make about controlling your children, your mate, and yourself.

Despite an uptight upbringing, you can reclaim the most vital parts of your life, emotions, and dreams that may have withered in childhood.

Despite a childhood in which you had little say, you can discover a new richness to your voice in the world.

Despite growing up with unhealthy family ties, you can fashion more nourishing relationships with those close to you.

Despite your own painful childhood, you can significantly increase the chances that your children will not suffer the pains you suffered.

Despite a troubled past with your parents, you can develop a more realistic and satisfying relationship with them as they near the end of their lives, and with their memories after they are gone.

It is possible to be yourself even if you had to be always "on" for your parents. It's possible to use your feelings for your betterment, not against it. No matter what your age or how restrictive your upbringing, it's possible to fulfill your personal promise and find the contentment that was derailed by parents who may not have known better or couldn't have done things any differently.

All these things are possible by achieving greater *individuation* from a controlling upbringing—and it begins with emotionally separating from the hurtful and problematic habits of your parents and family system. Individuation also includes setting right what was knocked out of balance by overcontrol and redefining yourself and your life in your own terms.

By individuating you can better know the hero or heroine in you: the biggest and strongest parts of you that helped you survive when you were smallest and weakest. Precisely because your parents were so controlling, you had to develop many strengths to survive—resourcefulness, intuition, perseverance, and sensitivity, for example. Luckily, the skills you taught yourself in navigating a difficult childhood are yours to keep and can be quite useful in adulthood. You deserve to feel independent and whole, to have healthy boundaries, to have free speech and open emotional expression. You deserve to heal.

How This Book Is Organized

This book is organized in three parts:

Part One, "Naming the Problem," will help you see the full extent of parental control by describing in detail eight styles of controlling parents. You'll be able to determine which of these types—or combination of types—fits one or both of your parents. When you know your parents' styles, you can better recognize the continuing effects of their early control on you.

Part Two, "Understanding the Problem," will help you reckon with control's lingering costs. You'll begin to understand the complex, powerful process of overcontrol and find answers to major quandaries such as, "How did my parents do it?" and "Why do I feel the way I do?" You'll gain clarity on your feelings as a child and discover connections between those feelings and your present-day problems. By exploring the aspects of yourself you had to disown or distort in childhood, you'll pave the way for reclaiming your total self. And, you'll get a clear sense of why your parents acted as they did, which will hasten your healing.

Part Three, "Solving the Problem," helps you let go of a painful childhood and the lasting effects of unhealthy control so that you can *emotionally leave home*. We'll explore a broad array of paths to healing, along with exercises you may find helpful. This section will help you design your own healing process, at your own speed, in a way that suits you best.

PART ONE

◆

Naming the
Problem

1

HEALTHIER PARENTING VERSUS CONTROLLING PARENTING

If you bungle raising your children, nothing else matters much in life.

—JACQUELINE KENNEDY ONASSIS

Healthy parenting is simple: Raise children well and set them free. Being a healthy child is also simple: Play, learn, grow up, and leave home.

But while both job descriptions are simple, neither is easy. The primary difference between healthier families and controlling families is that the parents in healthier families allow their children to grow up as persons in their own right.

Controlling parents fail to protect and nurture, robbing their children of playtime by using harsh or erratic discipline. They model unhealthy habits and hamstring their sons' and daughters' efforts to individuate. That's why people who grow up controlled sometimes struggle to emotionally leave home well into their thirties, forties, or fifties.

The following chart shows eight major differences between healthier families and controlling families. You might notice which side of the chart most closely parallels your childhood experience.

Characteristics of Healthier Vs. Controlling Families

Healthier Families	Controlling Families
1. **Nurturing Love** • Parental love is relatively constant • Children get affection, attention, and nurturing touch • Children are told they are wanted and loved	1. **Conditional Love** • Parental love is given as a reward but withdrawn as punishment • Parents feel their children "owe" them • Children have to "earn" parental love
2. **Respect** • Children are seen and valued for who they are • Children's choices are accepted	2. **Disrespect** • Children are treated as parental property • Parents use children to satisfy parental needs
3. **Open Communication** • Expressing honest thought is valued more than saying something a certain way • Questioning and dissent are allowed • Problems are acknowledged and addressed	3. **Stifled Speech** • Communication is hampered by rules like "Don't ask why" and "Don't say no" • Questioning and dissent are discouraged • Problems are ignored or denied
4. **Emotional Freedom** • It's okay to feel sadness, fear, anger and joy • Feelings are accepted as natural	4. **Emotional Intolerance** • Strong emotions are discouraged or blocked • Feelings are considered dangerous

Healthier Families	Controlling Families
5. **Encouragement** • Children's potentials are encouraged • Children are praised when they succeed and given compassion when they fail	5. **Ridicule** • Children feel on trial • Children are criticized more than praised
6. **Consistent Parenting** • Parents set appropriate, consistent limits • Parents see their role as guides • Parents allow children reasonable control over their own bodies and activities	6. **Dogmatic or Chaotic Parenting** • Discipline is often harsh and inflexible • Parents see their role as bosses • Parents accord children little privacy
7. **Encouragement of an Inner Life** • Children learn compassion for themselves • Parents communicate their values but allow children to develop their own values • Learning, humor, growth and play are present	7. **Denial of an Inner Life** • Children don't learn compassion for themselves • Being right is more important than learning or being curious • Family atmosphere feels stilted or chaotic
8. **Social Connections** • Connections with others are fostered • Parents pass on a broader vision of responsibility to others and to society	8. **Social Dysfunction** • Few genuine connections exist with outsiders • Children are told "Everyone's out to get you" • Relationships are driven by approval-seeking

The Consequences of Unhealthy Parenting

Healthier parents try, often intuitively and within whatever limits they face, to provide nurturing love, respect, communication, emotional freedom, consistency, encouragement of an inner life, and social connections. By and large they succeed—not all the time, perhaps not even most of the time, but often enough to compensate for normal parental mistakes and difficulties.

Overcontrol, in contrast, throws young lives out of balance: Conditional love, disrespect, stifled speech, emotional intolerance, ridicule, dogmatic parenting, denial of an inner life, and social dysfunction take a cumulative toll.

Controlling families are particularly difficult for sensitive children, who experience emotional blows and limits on their freedom especially acutely. Sensitive children also tend to blame themselves for family problems.

The more your experience mirrored the "Controlling Families" side of the preceding chart, the greater your risk of inheriting distorted views. You might note whether one or more of the following five distortions causes problems in your present life:

1. Distortions of Power and Size

If one or both parents demanded absolute control and dependence or treated you in ways that made you feel small, you may have inherited distortions of power and size. You may automatically view yourself as less capable than others or, alternatively, as so big and powerful that you have to protect others from yourself. You may feel you lack permission to do things that are within your perfect right. You may feel intimidated or, conversely, contemptuous in the presence of authority figures. Distortions of power and size can handicap you at work, as a parent, and in your other intimate relationships.

2. Distortions of Feeling and Wanting

If emotions were banned, inflated, or feared, and your desires shamed or thwarted, you may have inherited distortions of feeling and wanting. You may regard emotions such as anger, fear, sadness—even joy—as life-threatening and overreact to them. You may be unable to tolerate a loved one's strong feelings. You may deprive yourself of legitimate yearnings or live with unrealistic hopes. You may unconsciously expect life to be painful and, as a result, you may automatically become uncomfortable whenever good things happen. Distortions of feeling can lead you to fear or ignore your emotions and misinterpret the emotions of others. Distortions of wanting can leave you feeling deprived.

3. Distortions of Thinking

If truths were denied, perceptions discounted, or blame and shame heaped on you, you may have inherited distortions of thinking. You may accept overcontrol from others, thinking that it is normal. You may chronically doubt your perceptions. You may leap to conclusions based on all-or-nothing reasoning. Distortions of thinking may lead you to avoid personal responsibility or to assume too much responsibility for others' actions. Distortions of thinking can put you at risk for misreading others and yourself.

4. Distortions of Relating

If closeness was dangerous, or if you were infantilized for too long, or if you were thrust into the caretaker role too soon, you may have inherited distortions of relating. You may be unable to get close to others even when you want to. You may unwisely trust others or be unable to trust at all. You may see others as threats or as saviors—not simply as people. Distortions of relating can rob you of intimacy and pleasure.

5. Distortions of Self and Identity

If your intuition, initiative, or needs were devalued, you may have inherited distortions of self and identity. You may underrate your abilities, undercut your potential, or underplay your strengths. You may banish parts of your personality, present a false front to others, or see yourself as an object instead of a person. Distortions of self leave your primary relationship—that with yourself—underfueled.

But remember: *Knowledge is power.* By recognizing these distortions in your life, you can heal them.

How You Responded to Overcontrol

While you had relatively little power as a child, you were not simply passive. You were a growing, coping being who did your best to survive. Controlled children generally seek one or more of the following coping strategies:

1. **Complying** by doing what parents want

2. **Rebelling** by opposing parental wishes

3. **Distracting** through clowning around or emotional outbursts

4. **Dissociating** by numbing out, escaping into addiction, or becoming virtually invisible

5. **Outdoing** by trying to gain parental favor or by being more perfect than their parents

Each of these strategies has both payoffs and costs:

- Compliant children avoid some parental wrath but may forfeit autonomy.

- Rebellious children gain autonomy but may adopt a negative self-image.

- Distracting children avoid negative attention but may lose stature.

- Dissociating children escape control but may lose a sense of self.

- Outdoing children gain parental approval but may internalize unhealthy values.

Continued into adulthood, these coping strategies carry with them both assets and liabilities. For example:

Complying: The ability and willingness to know and meet people's needs and wishes can be valuable assets. But do you comply with others' demands even when it's not in your best interests?

Rebelling: An independent, free spirit is a rare gift. But do you automatically rebel even though you lose more than you gain?

Distracting: Being able to lighten up a situation is a special and needed talent. But do you find yourself distracting when it would be more helpful to face a situation head-on?

Dissociating: Being able to turn inward and shut out stress can help in concentrating and relaxing. But are there times when dissociating makes unhealthy situations worse?

Outdoing: Self-discipline and the drive to accomplish are powerful assets. But do your efforts sometimes feel compulsive or involuntary?

As an adult, you have choices beyond these five coping strategies. Although they helped you survive the emotional minefields of your family, they were reactive. This book will show you how to create a healthier balance of power between you and your parents, both your actual embodied parents and your internalized parents—those inner critics who shape and shame. A healthier balance of power can help you fashion proactive rather than reactive coping strategies. For example, the following new directions may bring growth:

If you tend to comply, you may want to find more opportunities to go your own way and make independent choices.

If you tend to rebel, you may want to strengthen your ability to more readily accept unpleasant (though not abusive or destructive) people or situations.

If you tend to distract, you may want to practice sticking with and observing uncomfortable (though not abusive or destructive) situations.

If you tend to dissociate, you may want to work harder to focus on the here and now.

If you tend to outdo, you may want to find new ways to ease up and relax.

Controlling Family Loyalties: Ties That Blind

People find it hardest to recognize problems in their own families.

—TIPPER GORE

As you read this book, feelings of family loyalties may become triggered. Such feelings, however uncomfortable, are perfectly normal.

Healing from growing up controlled can be hard work. It's troubling to acknowledge shortcomings in your parents and yourself. It's painful to conclude that, if not for your parents' limitations, you might have grown up happier, with healthier relationships and a less troubled life. Guilt, anger, fear, sadness, and love make relationships with our parents among the most complicated in our lives.

The loyalties and inhibitions installed during your preverbal years can make it hard to explore your upbringing, even years later, as an adult. Many of us grew up in black-and-white, all-or-nothing families. The result, black-and-white, all-or-nothing thinking, can be a form of denial, which exists to keep hurt away, at least temporarily. When you start to explore past pains and current problems, you break that denial. In so doing, you may feel sad, mad, disloyal, exhilarated, lonely, and free, in quick succession or simultaneously.

Separating from our parents in order to define our own identities is the chief task of adolescence. It's no surprise that the teenage years are so tumultuous; that goes with the individuation territory. In controlling families, however, individuation rarely takes place in the teens because controlling parents tend to hold on too tightly or push too hard. Many controlled teenagers feel too loyal, confused, afraid, or wounded to make the break. The good news is that individuation in our twenties, thirties, forties, or fifties takes place on a deeper level, with more balance and greater growth, than is possible during adolescence.

As you read on, you might notice your inner dialogue. Thoughts like "Don't blame others" or "It's all in the past" may actually be the internalized voices of your parents. These thoughts may feel like warnings to stop exploring, but they offer you valuable information. Observing these messages can show how voices from your past reach into your present to dictate behavior. If you occasionally feel awash in "wrong" or conflicting feelings, questions, or insights, I suggest that you're not doing something "wrong"—you're making progress. By exploring the paradoxes of your feelings and your relationship with your parents you are embracing more than the either-ors you grew up with. You are gaining freedom from overcontrol.

Top 10 Guilt-Inducing Family-Loyalty Thoughts

Several concerns commonly occur at various stages of individuating and healing among adults who grew up controlled. For readers who may feel ambivalent about revisiting their family's control, the following ten concerns and responses may help you sort through your feelings and decide how deeply you want to explore.

If you're ready to plunge forward, skip this list. If at some point later on you feel bogged down in your growth and healing, that's the time to refer back to this section.

1. *"I owe my parents respect, loyalty, and gratitude. They made a lot of sacrifices for me and I wouldn't be here if not for them."*
   Confronting what was unhealthy in your upbringing doesn't make you disloyal to your parents, and it doesn't indicate that you're downplaying their contributions. Rather, it means you're being loyal to yourself. There's nothing disrespectful about asking honest questions when they're in your own best interests. If you came from a controlling family, developing a flexible sense of family loyalties—that doesn't diminish your sense of yourself or exist

in all-or-nothing terms—can allow you to see both the good and the bad in your past and in your parents. It's both helpful and healing to study how unhealthy loyalties may have been instilled in you and whether you are trapped by them even today.

2. *"What if exploring this makes me feel anger, pain, fear, or grief?"*

You don't have to explore your childhood. It's never easy, particularly if your childhood wasn't easy. Yet it can be freeing.

As you delve into your past, emotions can be intense because they often include leftover emotions you couldn't fully experience as a child. If you had controlling parents, they were probably terrified of being overwhelmed by feelings. That's a major reason for why people control. If your parents feared feelings, they probably tried to avoid, alter, or block all family members' emotional expressions.

Reclaiming your independence may mean connecting with anger, sadness, hurt, rage, loneliness, desolation, or anxiety. As strong as these feelings are, they will eventually pass. By examining and embracing your feelings, you strengthen emotional muscles that were underused in childhood.

3. *"It's all in the past, so what good does it do to go over it?"*

While exploring a painful childhood can initially seem to make your life more difficult, it will eventually help you to enjoy a healthier present and future. Your sense of self can change. Your relationship with your parents can change. Your willingness to be yourself despite others' disapproval can change.

For many years I downplayed my parents' influence on me. Looking back, I can see why: It was painful to admit that they had let me down, even if unintentionally; it hurt to face my desperate attempts to be accepted, hiding my needs and weaknesses, yet still never feeling accepted; and, most of all, it grieved me to realize that I, like all children, was powerless to stop my parents from hurting me.

It can be hard to accept the idea that parents have so much of an impact on us. It may be hard to remember that as children we were relatively helpless and dependent. It can be terrifying to admit that your parents muffed one of the biggest jobs of their lives—raising you. It can be so threatening, in fact, that many of us tend to rationalize away that hurt. Freedom lies in seeking a balanced view that neither minimizes nor overstates.

4. *"What if my parents die before I sort all this out?"*

Watching a parent age and die is always tremendously difficult. If your parents were abusive or controlling, their aging can bring a special set of emotional challenges.

Few of us have "finished" relationships with the dead. It can feel devastating if a parent dies before you have had a chance to say your piece or make your peace. But you can still say what you have to say in a letter, meditation, or poem even after a parent is gone. Part Three, "Solving the Problem," will offer help in coping with the aging and death of controlling parents.

5. *"It wasn't that bad. Lots of other children had it much worse."*

One forty-year-old woman whom I interviewed told me, "My parents never hit me and they certainly gave me food and shelter and an education, so I guess I don't have much to complain about compared to children who were hit or molested." Yet she was ruthlessly controlled by her parents. The pain of emotional maltreatment can be as deep and long-lasting as that of physical abuse. A slap, shove, insult, or look all hurt equally deeply. In fact, many people who were physically abused say it was the words, not the blows, that hurt most.

You don't have to be hit or molested or left without food or clothes to be left with the effects of long-term abuse. Overcontrol, neglect, and cruelty are all painful—and all wrong. I'll be sharing stories from a wide variety of difficult childhoods in the hope that you will find, rather than invalidate, yourself.

6. *"I don't want to be a victim and blame others for my problems."*

Self-help books and groups are criticized for turning us into a nation of "whiners" who blame others for our own issues and take no responsibility for seeking remedies. To be sure, some people do get stuck in the "victim" stance. Yet in my experience, most people read self-help books or participate in self-help groups because they care about the quality of their lives. In working with women and men who grew up controlled, I've found that most have trouble blaming anybody *but* themselves because they tend to accept their parents' points of view at the expense of their own.

Children of controlling families aren't trained to act in their own best interests; they're trained to serve and take care of their parents. Questioning your parenting and discovering connections between your current problems and your upbringing is acting in your own best interests—although initially it may feel awkward.

It's important to remember that even if your parents loved you, their control cost you a great deal. This book is not about blaming parents for their mistakes, but it is about understanding their mistakes so you no longer suffer the consequences.

7. *"My parents were only doing what they thought was right. Better to forgive and forget."*

Many controlling parents do what they think is right, but it doesn't mean it was right for you. Vengeance is the last thing this book is about. However, in my experience, forgiving another's transgressions before you're ready to can be as destructive as vengeance. In Part Three we'll explore forgiveness in depth.

8. *"I don't remember much of my childhood. How do I know if I was controlled?"*

It's not specific memories of childhood experiences that need to be healed. Rather, it's the emotional experience of growing up controlled and the decisions you may have unwittingly carried into your adult life that do the harm. Powerful, unseen injunctions often serve as barriers to seeing your past for what it was. Matching your experiences against those of the people you'll read about can help clarify your past.

9. *"My parents, family, or friends might ridicule or reject me if I explore this."*

A message like this is a signal that even the thought of others' disapproval has the power to stop you cold. Yes, some might disapprove. But a big part of individuation is seeking your truth even when others disagree. If you have parents or friends who attack you for striking out in your own best interests, exploring those relationships may be all the more compelling.

Investigating your past can be as private a process as you choose. It's possible to let go of past limits and achieve new freedom with your parents without saying a single word to them. Lots of people have.

10. *"It's hopeless to think that I can change, given how long I have been this way. It's hopeless to think my parents will ever change."*

This statement reflects the perfectionistic, all-or-nothing thinking common in controlling families. Psychological change can be difficult and slow, but it is not all or nothing. While your parents may never change, your healing is not dependent on what they do. Your healing depends on what you do. Even a minor adjustment in your feelings, behaviors, and relationships can bring you huge payoffs.

Identifying Your Parents' Styles

Nearly all controlling parents embody one or more of the eight "styles" of controlling parenting. These styles provide a "You Are Here" point on the map of unhealthy control.

Identifying your parents' styles can help you make sense of what didn't jibe in your family. Remember the series of lenses an eye doctor alternates before your eyes until you find the ones that enable you to see most clearly? Recognizing your parents' styles offers the right lens to bring into focus the underlying values and themes with which you were raised. The more clearly you view your family's themes, the more readily you can become your own person.

You may find elements of one or more of these styles present in either or both of your parents:

- **Smothering**. Terrified of feeling alone, Smothering parents emotionally engulf their children. Their overbearing presence discourages independence and cultivates a tyranny of repetition in their children's identities, thoughts, and feelings.

- **Depriving**. Convinced that they will never get enough of what they need, Depriving parents withhold attention and encouragement from their children. They love conditionally, giving affection when a child pleases them, withdrawing it when displeased.

- **Perfectionistic**. Paranoid about flaws, Perfectionistic parents drive their children to be the best and the brightest. These parents fixate on order, prestige, power, and/or perfect appearances.

- **Cultlike**. Distressed by uncertainty, Cultlike parents have to be "in the know," and often gravitate to military, religious, social, or corporate institutions or philosophies that allow them to feel special and certain. They raise their children according to rigid rules and roles.

- **Chaotic**. Caught up in an internal cyclone of instability and confusion, Chaotic parents tend toward mercurial moods, radically inconsistent discipline, and bewildering communication.

- **Using**. Determined never to lose or feel one down, Using parents feed off their children emotionally. Hypersensitive and self-centered, Using parents see others' gains as their loss, and consequently belittle their children.

- **Abusing**. Perched atop a volcano of resentment, Abusing parents

verbally or emotionally bully—or physically or sexually abuse—
their children. When they're enraged, Abusing parents view their
children as threats and treat them accordingly.

- **Childlike**. Feeling incapable or needy, Childlike parents offer their
children little protection. Childlike parents, woefully uncomfortable with themselves, encourage their children to take care of
them, thereby controlling through role reversal.

Of course, most controlling parents are a combination of styles,
with one, two, or three predominating. My father, for example, was a
Perfectionistic-Cultlike-Using parent. And, as you will shortly discover, certain style combinations tend to go together—both within an
individual parent as well as between controllers who marry.

Next: Portraits

The next chapters contain portraits of adults who grew up with
parents having at least one of these eight styles. By matching your
experiences against theirs, you can see your family's early atmosphere
more lucidly.

Recognizing these controlling styles can also help you to identify
your internalized parents—the "inner critics" we all carry in our heads.
In Part Two we'll revisit these eight styles and use them to help you
free yourself from your inner critics.

You might also notice whether one or more of these eight styles
strikes a chord in the way you, in your worst moments, relate to others. Recognizing these unwelcome inheritances will enable you to dis-
mantle those emotional land mines left over from early control.

2

SMOTHERING PARENTING
Life Under a Microscope

You love me so much, you want to put me in your pocket. And I should die there smothered.

—D. H. LAWRENCE

Key Characteristics of Smothering Parents:

- Control through overbearing scrutiny

- Fear rejection or being alone

- Cannot differentiate between their own wants and those of their children

- Discourage their children's individuality

Potential Consequences of a Smothering Upbringing:

- Lack of healthy interpersonal boundaries

- Difficulty with intimacy and commitment

- Intense dependency

- Poor body image

- Reduced initiative

Slight, porcelain-skinned Margaret, a thirty-three-year-old attorney specializing in family law, grew up with a lawyer father who loved heated discussions, always insisting that Margaret argue with him and defend her positions. Unfortunately, he never allowed her to win, badgering her until she capitulated.

At age nine, Margaret began reading a book about a veterinarian, which her father covertly confiscated since he wanted her to be a doctor, not a vet. When Margaret asked where the book had gone, her father responded with, "What book?" When she was twelve, Margaret developed a taste for bland foods—vanilla ice cream, white bread, and potatoes—so her father endlessly shoved the spicy foods he preferred under her nose. As sixteen-year-old Margaret was writing her college application essays, her father grabbed them, read them disapprovingly, sat down at the kitchen table, and rewrote them. When seventeen-year-old Margaret was packing for college, her father began yanking clothes out of her suitcase, telling her exactly what and how to pack.

Feeling overscrutinized is the hallmark of growing up with a Smothering parent. While Smothering parents can seem incredibly caring, their form of love can breed unhealthy dependence. The endless attention they bestow has its price, for when a Smothering parent is always there for the child, the unspoken agreement is that the child will always be there for the parent.

Smothering parents seem unable to see their children as separate human beings but, rather, see them as worlds to be controlled. Therapists call Smothering households "enmeshed," full of what family therapy pioneer Murray Bowen called emotional "stuck-togetherness" or what author Jane Middleton-Moz has termed emotional "superglue." At their core, Smothering parents cannot let their children be independent because it reminds them of the fact that their children will eventually grow up and pursue their own lives. This prospect leaves many in-your-face parents feeling abandoned and invalidated.

Margaret recalls, "My father had this uncanny way of questioning, of saying, 'Are you sure? Are you sure?' You couldn't say no. After all, as a lawyer he convinced people for a living." Margaret coped by mentally replaying conversations with her father every night. "I'd lay in bed and tell him off, telling him that this was counterproductive to my growing up. But I would never dissent out loud."

Margaret acknowledges that her father, who had Perfectionistic as well as Smothering characteristics, may have had good intentions. But his heavy-handed actions, like rewriting her college application essays or repacking her bags, left her feeling like her "feet had been chopped off."

Uniform Feelings

While some parents, like Margaret's dad, want their children to mimic their thoughts, others focus on uniformity of feelings.

Sharon, a thirty-one-year-old graduate student, grew up under the emotional thumb of her father, David, a Jewish Holocaust survivor. David was a newborn when his parents managed to get him to a Catholic orphanage just before they were sent to concentration camps. Miraculously, both parents survived, although they became estranged, and David's mother subsequently found him. For years David's parents passed him back and forth, even resorting to abduction. Perhaps as a result, David could not stand for his own child to be out of his sight. "Even when I was four or five he carried me around like I was a baby," Sharon says.

This intense attention was not always positive. In her adolescence Sharon's father called her "Buck Teeth" and told her, "Your thighs are as big as mine." He defended his remarks as "good character-building."

At sixteen, after her parents divorced, Sharon confided to her father that she was having a hard time adjusting to her new step-mother; her dad became furious and branded her "self-centered."

It wasn't until years later that Sharon realized her Smothering father was self-absorbed. "He has this chasm in him: people in black-ness, screaming, climbing walls, being gassed. I have it too. It is in our psyche. But because he has this pain inside him he thinks that nobody else's pain is as great. He could never hear me out when I felt hurt."

In the midst of a divorce when she was interviewed, Sharon had often picked controlling men as partners. "My heart just opens to men, like it did to my dad, and I get taken advantage of," she admits.

I do not minimize the trauma of the Holocaust or the legacy of emotional difficulty facing survivors and their descendants. That Sharon is suffering the impact of the Holocaust a half century later is testimony to the enduring potency of that trauma. For Sharon's father, like many other Holocaust survivors, generating life through child rais-ing became a sacred pursuit. As one Holocaust survivor told *The New York Times* on the fiftieth anniversary of the death camp liberations, "Our vengeance was rebuilding life" through children.

When Sharon's father sought to "rebuild life," the uncertainty and horror of his crucial early years, as well as his parents' ongoing rift, almost certainly led him to focus intently on his daughter in an effort to protect her. Perhaps David, having lived as an orphan until age four, was also unconsciously trying to live out through his daughter the childhood he had never had. Unfortunately, his grip was too tight for Sharon's optimal emotional development.

Uniform Siblings

While some Smothering parents control their children's thoughts or feelings, others make their children conform to each other. Boys and girls in many large but non-Smothering families sometimes feel they're just faces in the crowd. Smothering families take this conformity to extremes.

Colleen, a thirty-three-year-old graduate student, is the oldest of seven children from an Irish-Catholic family. She vividly recalls being fourteen and sitting in her assigned seat at the family dinner table. Her father would signal each child in turn to report on what had happened at school that day. If one refused, none could talk. Though her parents may have had egalitarian aims, Colleen felt trapped in a tyranny of sameness. On family car outings, when any of the children misbehaved, her father hit whoever was closest, whether that child had misbehaved or not. If the innocent victim protested, Colleen's father would say, "Who cares? You all need to behave."

This conformity left Colleen feeling devalued. "Who we were wasn't important. All that mattered was how we fit into the family," Colleen said. "We were all just nobody special."

Yet in ways never acknowledged by her Smothering parents, some differences were allowed, and even fostered. Her brothers' activities received more attention than Colleen's or her sisters': "My parents went to my brothers' hockey games but never came to my basketball games. I felt like one of the things most snuffed out was my feminine side. My parents acted like the only way to be valuable was to be like a man."

In her adult life Colleen has found it difficult to express her feelings or honor her intuition. She has repeatedly found herself in struggles with authority figures and has been unable to sustain a long-term intimate relationship.

Uniform Values

Some Smothering parents become overbearing in encouraging their children to adopt their values.

At age six, Cui, now a twenty-seven-year-old sales representative, lay in bed reciting her multiplication tables as her father stood over her. This nightly ritual was part of her immigrant Chinese parents' campaign to stress academic achievement. During grade school, whenever Cui's mother's friends visited, her Smothering, Using mother would hustle Cui off to her room to retrieve her awards for academic excellence.

"Mom used me as a showcase for her friends," Cui recalls. "I was like the chess-champion daughter in The Joy Luck Club *who was forced to play and was valued only when she won. When I saw that movie, I started crying during the opening titles and didn't stop until after the closing credits."*

Cui was raised with the expectation that she would become a doctor. But when she told her parents during her sophomore year at Princeton that her premed grades weren't high enough, her parents were crestfallen. Within minutes her mother brightened and said, "Okay, then you'll be a lawyer!"

Smothering parents are unaware of how little they see their children as separate. They easily make incursions into their children's lives because they do not see their actions as intrusive. Cui says, "I got strokes for external accomplishments, never just for being the person I am. I feel like I have lived my life to please others."

Both Cui's parents came to the United States in their teens, cut off from their families, after narrowly escaping from China before the 1949 civil war broke out. Their struggle to fit into American culture is familiar to many immigrant and multi-ethnic families. In addition, in the Chinese culture great value is placed on academic achievement, and individual freedom and responsibility is viewed differently than it is in American culture. But in their efforts to adopt American values, Cui's parents lost sight of their daughter's needs.

Uniform Lifestyle

While Cui's parents pressured her to adopt certain values, other Smothering parents pressure their children *not* to, even when those values reflect what is deepest in a child's heart.

Sally, a thirty-five-year-old computer programmer, recalls the time when as a college sophomore she was pulling her Volkswagen beetle into her family's driveway during spring break. Her father came out to help her unload her luggage and, noticing a pink triangle bumper sticker, asked what it meant. As she carried in her suitcase, she told him that the pink triangle was a symbol of lesbian and gay liberation. Returning to continue unloading, Sally found her father scraping the sticker off her car, telling her, "That's not the kind of thing you want on your car. The kind of attention it will attract, you don't want."

"That was my coming out," recalls Sally, who'd known she was a lesbian since she was thirteen, but had not told her parents. Her dad has not mentioned her sexual orientation since the incident. "My father still gives me talks about dating and how I have to get out there to find a good man," Sally says, smiling.

Sally's father had intruded into her business all her life. He insisted she finish her vegetables, serving anything she didn't eat the next morning, cold, at breakfast; making her eat brussels sprouts was a regular punishment. He would wake Sally each morning, pulling the covers from her if she tried to sleep for a few extra minutes. "He was very invested in his family," Sally says. "It never felt malicious or malevolent, but we were not supposed to have independent wills. Disagreeing would have meant 'I don't love you' to him."

Excessive Scrutiny

Some Smothering parents scrutinize their children in the most invasive ways possible.

When forty-eight-year-old social worker Tina was four, her mother thought Tina was too thin, so she hovered over her at mealtimes until she'd finished the huge helpings prepared for her. But when Tina was six, her mother decided her daughter was too fat. She put her on a crash diet and, whenever Tina was outside the home, taped a sign to Tina's back reading, "Please Do Not Feed Me."

Her mother, a nurse, scrutinized Tina's bodily functions and provided frequent "home remedies": enemas if Tina had not had a daily bowel movement, douches as early as age nine, and penicillin shots stolen from the hospital at the first sign of Tina's having a cold or a sniffle. "Growing up was like being a patient in a sick ward," Tina admits.

Her mother picked out her daughter's clothes without consulting her. Tina recalls pictures of herself as a somber little girl with bangs and a turned-up nose, wearing garish outfits four sizes too big. "I was overfed, horribly dressed, had thick glasses, and was very nervous," Tina remembers. "I've seen people in wheelchairs, shrunken and paralyzed, who have a better body image than I had."

Like many controlled children, Tina had little privacy. She and her siblings were forbidden to close doors; they showered, used the toilet, and slept, all with the doors open. "It wasn't until I went away to college that I realized this was not normal."

Tina found that it was fruitless to be herself except in her private world of imaginary playmates and someday hopes. The smothering scrutiny in her childhood has translated into an adult feeling of "bottled-upness." For much of adulthood Tina has lacked confidence in her choices, expected others to think poorly of her, and found it hard to ask for what she wants.

To Smothering parents, a child's dissent means rejection, and a child's independence means "I don't love you." Margaret's mind, Sharon's feelings, Colleen's individuality, Cui's temperament, Sally's sexual orientation, and Tina's body were too separate for their parents to tolerate.

Self-Assessment

My parent(s):

- Overscrutinized my personal habits
- Did not tolerate differing viewpoints or tastes
- Had trouble coping with strong emotions
- Tried to dictate my career and life choices
- Seemed to have difficulty being alone

Next: Depriving Parenting

While Smothering parents engulf their children with too much attention, the next style of controlling parents, Depriving parents, do exactly the opposite. They emotionally abandon their children by withholding love, support, and attention, which gives them commanding power.

3

DEPRIVING PARENTING
Playing "Take Away"

When my mother was disappointed in me she'd become inaccessible and emit a silent scorn. I could have withered away to nothing. It felt like death.

—SAMANTHA, 40, AN ARTIST

Key Characteristics of Depriving Parents:

• Control through withdrawal, disapproval, or banishment

• Become emotionally unreachable

• View happiness and good fortune as scarce

• See love as a commodity to be withheld when the parent is displeased

Potential Consequences of a Depriving Upbringing:

• Self-doubt

• Depression

• Lowered expectations and confidence

• Feeling unloving and unlovable

• Slowed development of social skills

Forty-year-old artist Samantha, a pretty, curly-haired blond in the throes of puppy love at thirteen, blushed when her eighth-grade boyfriend gave her roses, her first. Rushing home, she breathlessly asked her mother for a vase. Instead, she got a sour look and an order to put the roses in the garage. For the next week, Samantha spent a good part of her days in the hot, dusty garage, watching her roses slowly wilt.

Samantha's mother also became downright sullen when Samantha wanted something. At seven, on a daylong shopping trip with her mother, Samantha asked to go to the bathroom and was told, "Shut up. If you ask one more time I'm going to leave you behind in this store."

By age fourteen Samantha had become interested in spirituality and asked to go to a Christian summer camp. Her mother informed her that she couldn't go because of a family vacation that had already been scheduled. Yet the camp date arrived and nobody went anywhere. "When my mother didn't want to do stuff she'd say we had a family-schedule conflict," Samantha remembers. "She just didn't want to inconvenience herself."

Conditional love is the trademark of Depriving parents. As long as their children conform to their desires, these parents lend emotional support. But when they're disappointed, Depriving parents withdraw their love—remarkably, instantly, and utterly. This leaves children so unsure of their standing that they're desperate to please. They learn that love is ephemeral and erratic or contingent upon good behavior. They also learn that they have little control over whether they receive love. Whereas Smothering parents tell their children what they *should* want, Depriving parents tell their children simply *not to* want.

Children who grow up with Depriving parents vividly recall the experience of repeatedly losing parental love and support. "My mother had so much more power over me because of what I didn't get than because of what I did get," Samantha confesses. "So I became even more dependent on her."

Perhaps Samantha's mother repeatedly deprived her daughter because she felt personally deprived. Perhaps seeing her daughter happy made her aware of her own unhappiness and she found the disparity intolerable. Since she didn't know how to make herself feel better, she may have unconsciously tried to make her daughter feel worse.

Samantha remembers how her Depriving, Abusing mother's approval washed in and out, like the tide. She frequently threatened to disown Samantha if she didn't follow parental desires; and, in fact, Samantha's twenty-four-year-old brother was disowned for dating a Native American against parental wishes. As soon as the couple broke up, Samantha's parents began speaking to their son again. And, when Samantha's nineteen-year-old sister announced that she was moving in with a boyfriend, her parents again threatened to cut ties. The sister backed down and lived at home for two more years until she married.

After Samantha graduated from high school, her parents commanded her to stay at home and make plans to become a nurse so she could take care of them in their old age. Instead, she left the state. Her parents did not speak to her for six years. Samantha finally returned for a visit, only to find that her room had been turned into a sitting room and her belongings thrown out. Her mother housed her in a trailer in the driveway, though a spare bedroom sat unused.

However, the worst deprivation Samantha suffered was the lack of parental protection: From ages nine to twelve, Samantha was repeatedly left with a grandfather who molested her.

Discounted Dreams

One of the major ways in which Depriving parents wound their children is by ignoring or discounting their future dreams.

David, a fifty-year-old highly successful salesman, was an only child of Jewish merchants in a small Mississippi town. He vividly recalls his elation when an aunt gave him a Brownie camera for his thirteenth birthday. David passionately wanted to be a photographer when he grew up, but his parents pronounced his career dreams impractical. "Photography is a waste of time," they said. "Stick to what is familiar and take over our store when we retire." To discourage his interest in photography, they refused to let him buy film. For months David pointed and clicked his useless camera for hours on end, eventually giving up.

Children need affection, encouragement, and physical contact. When they're deprived of them, they can feel invisible. David cannot recall being hugged, kissed, or told he was loved by his parents. The only physical comfort he had was from "Mammy," an African-American housekeeper who recognized David's needs and provided solace. On Saturdays she'd take him to a movie, where she was allowed to sit with him in the whites' section. He is convinced that, "If I hadn't had Mammy, I would have been in much worse shape."

David longed for recognition for his good grades, but his Depriving, Perfectionistic parents rarely made even a comment. "I did everything I was supposed to but they never approved. They never asked me how I felt, they just told me how I should react. Rules were more important than feelings." On family car rides, David's parents plunked him in the backseat and talked about him as if he weren't there.

While his mom kept after David to follow the family rules, his dad rarely spoke to his son. David remembers weekly drives to the big city with his father to take Hebrew classes for his bar mitzvah. During the two-hour trips not a word was spoken. While he felt thankful for his father's presence, he hungered for deeper emotional contact.

Looking back, David recognizes that there were some unavoidable reasons for the deprivation in his childhood. Both David's parents lost their mothers before age five. Because of those early losses, sadness lingered in his family for two generations. In addition, as Jews in the Deep South in the 1940s and 1950s, his family may have felt they needed to behave in a certain fashion. This may explain why David was told so often to be quiet and "proper."

Regardless of the reasons, David paid a price. Successful he may be, but he still struggles with loneliness. Lacking an early experience of steady emotional warmth, he has yet to have a long-term intimate relationship. "If I'd gotten a few hugs and a few moments of conversation in my childhood, it might have changed a few things," he muses. He can hardly bear to watch TV or movie scenes of fathers and sons. "It cuts me like a knife to see a father and son being close," he admits.

One of David's few joys comes on vacations, during which he takes photographs—in part, to replace the photographs he couldn't take as a child because he wasn't allowed to have film.

Ignored Gifts

While some Depriving parents, like David's, discount their children's interests, others ignore their children's innate gifts.

Shari is a thirty-two-year-old marketing executive who'll never forget fidgeting on the stage at her junior high school when she was thirteen, scanning the audience for her mother. Shari had placed in the 98th percentile on national scholastic tests and was being honored in an awards ceremony. Her mother never showed up.

"My mother's attitude was that it was expected of me to do well," Shari comments wryly. "The most she'd ever say was, 'That's nice, dear.' But if I didn't mop the floor perfectly, I'd get spanked."

In grade school her mother tried to force left-handed Shari to write with her right hand so she would be more "normal." When Shari asked "Why?" her mother taped her mouth shut.

Over time, Shari's straight-A average fell. At seventeen, she dropped out of high school and got a high school equivalency diploma so she

could work. "I don't remember getting any sort of positive direction from
my mother. She never told me what to do with my life. She just told me
what not to do."

Shari was raised by a Depriving, Abusing, single mother. Now herself a single mother, she recognizes that much of her mother's control
came from the demands of that daunting role. In addition, Shari, an
African American, believes that some of her mother's harsh control had
cultural and historical roots: "Black children grow up under a microscope. I think it goes way back to the slave days, when a black child
could have been killed for acting too rambunctious around white people. I think black parents from my mother's generation and earlier felt a
need to control their children so they wouldn't get negative attention."

Regardless of what Shari the adult can see in retrospect, Shari the
child felt unwanted. At nine, Shari was so upset she decided to drown
herself in the bathtub. When her mother left for work, Shari wrote a
will giving her toys to friends. She left the will on the kitchen table,
filled the tub, and stuck her head in. "Of course my attempt didn't
work," Shari says, smiling. "I kept coming up for air. Finally I just went
to bed and fell asleep." When her mother came home and discovered
the will in the kitchen, "She didn't seem concerned, just sort of sarcastic. My mother wasn't really there emotionally even when she was
there physically."

Shari's relationship with her mother has improved since Shari
became a parent. But she still wonders what she might have become
had her mother encouraged her: "I feel as if I have spent so much of my
life just trying to heal from my childhood. Who knows what I could
have been? I had the grades and the intelligence for even medical
school, if only I'd had more support."

Emptiness

Most parents of today's baby boomers grew up during the Great
Depression of the 1930s. Others, particularly recent immigrants, faced
poverty equal to or worse than that of the Depression. These experiences often saddled them with a deprivation mentality they never
seemed able to shake. As a result, some raised their children with a pervasive sense of emptiness.

Fifty-seven-year-old homemaker Roberta is thinking of sixth-grader
Roberta walking slowly around the block of the Philadelphia

neighborhood shoe store for the sixth time—embarrassed to return a pair
of patent-leather shoes.

Earlier that day, her mother had sent her to the store with ten dollars
to get new shoes. Roberta was captivated by a $9.49 patent-leather pair,
yet, seeing them, her mother began screaming that they cost too much.
Roberta meekly replied that her mother hadn't told her how much she
should spend, but her mother angrily ordered Roberta to return them.
After seven circuits around the block, Roberta mustered up her courage
and drifted inside, telling the clerk, "I can't keep these. My mother says
they cost too much."

Despite her family's relative financial security, every aspect of
domestic life dwelled in the shadow of scarcity and regimentation: "We
ate at six, eleven, and five and never deviated. Monday, Wednesday, and
Friday mornings we had cereal, milk, and half a glass of orange juice.
Tuesdays and Thursdays we had a poached egg. Saturday was
scrambled eggs. Sunday Dad made pancakes." Every week. Every year.
No exceptions. Snacking was not allowed between meals. Roberta wasn't
even allowed to look in the refrigerator.

With no allowance or job, eighth-grader Roberta began seeking
ways to hoard money. Just before lunch, she'd go to the library and
read something gory or upsetting—a regular was a description of
Hiroshima after the atomic bomb—so she would lose her appetite and
be able to keep the ten cents of lunch money.

Her childhood was barbaric. Her mother, distressed by changing
diapers, toilet-trained her daughter at ten months of age. Then she
took away daytime naps so Roberta would fall asleep for the night by
six P.M. Roberta remembers lying in bed as a toddler struggling to keep
her eyes open, humming Perry Como tunes, hoping she could stay
awake until her father came home to kiss her good night. Most nights
she didn't make it. On the rare nights that she kept herself awake long
enough, her father often gave her only a quick kiss before rushing
away, telling her, "Mom has dinner ready."

Roberta now struggles with alcoholism and eating disorders.
Unable to shake her parents' "severe Depression-era mentality," she
hoards food, eating only a bit of each meal, saving some just in case. For
our interview, Roberta asked that we meet at a local bakery yet arrived
with muffins of her own that she nibbled parsimoniously during our
talk: "I'm still hoarding. My parents controlled me. I controlled my
daughters. I control myself. I only wish my mother had once said,
'Keep it. Splurge.'"

Why Parents Deprive

Parents withhold affection for many reasons. Some, with severely limited access to their feelings, have little love to give. Many controlling parents are emotionally disturbed or self-absorbed and can barely perceive others' troubles. Still others are so uncomfortable with touch and intimacy that they cannot allow their children near them.

Parental deprivation can be deliberate, but more often it is instinctive, unwitting, or unconscious. It's what the parents learned. It's how they were raised. Depriving parents often think they're toughening up their children to survive a hard, cruel world. For them, life is about prevailing in a hostile environment; emotional caretaking and closeness are low priorities. Other Depriving parents try to enforce a Puritan ethic. Telling their children not to hope for much may reflect the way in which the parents themselves deal with disappointment. They believe that if you don't hope for much, you won't be as disappointed when the inevitable losses of life occur.

Strikingly, more than half of those people interviewed, not just those raised by Depriving parents, wondered as children if their parents really loved them. Such deprivation can make children feel so unworthy that by the time they reach adulthood they expect abandonment in relationships or find commitment terrifying. They may become hypersensitive to signs that a partner might leave.

Self-Assessment

My parent(s):

- Played "take away" with their love and approval
- Gave me scarce praise or physical affection
- Threatened to disown or disinherit me
- Viewed good fortune as scarce or unattainable
- Seemed cold or unfeeling

Next: Perfectionistic Parenting

The next style of parents, Perfectionistic parents, endlessly drive themselves and their children to be the "best."

4

PERFECTIONISTIC PARENTING
A Place for Everything (And Everything Had Better Be in Place)

We find fault with perfection itself.

—BLAISE PASCAL

Key Characteristics of Perfectionistic Parents:

- Control through pressure to be perfect and the best
- Mortally afraid of flaws, disorder, or uncleanliness
- Driven and compulsive
- Emphasize appearances, status, material goods, or what others think

Potential Consequences of a Perfectionistic Upbringing:

- Emotional "bottled-upness"
- Feeling valued for what you *do* instead of who you *are*
- Compulsivity
- Second-guessing and self-doubting
- Depression

Twenty-eight-year-old Will is a teacher, but thirteen-year-old Will, his sandy hair cropped close for speed in swimming, was a sure bet to make his school's junior varsity swim team. His father was insisting that Will bypass junior varsity and try out for varsity two years earlier than the normal age. On tryout day his dad drove him to school, unleashing a barrage of pressure and coercion. During tryouts, Will failed to make the team. Afterward, he recalled, "I got a two-hour lecture, with me in tears being grilled about why I let that happen."

The next year Will did make varsity and became a champion swimmer. But when he barely missed qualifying for the nationals, his dad berated him for his shortcomings rather than acknowledging what he had accomplished.

Will's father bore a trademark of Perfectionistic parents: the conviction that life is a performance and that anything short of perfection is failure. While most parents want good things to happen to their children, Perfectionistic parents insist that their children *make* good things happen: Will is convinced that, "My father's life was filled with regrets. He felt he hadn't become enough. He was trying to beat the game of life through me."

As a child, Will played his father's game devotedly: "I'd start out every school year saying, 'I'm going to do it perfectly, twenty-six hours a day, and make them proud.'" By his teens, Will felt numb. "I had nobody I could talk to. I didn't know what I wanted or what I felt. I remember looking at the mirror and feeling that the reflection had my soul."

As an adult, Will has had problems with authority figures and has carried a hard emotional edge that has nearly precluded intimate relationships. A combination of twelve-step programs and martial arts has helped him begin to recover his soul from the mirror.

Pressure to Perform

Perfectionistic parents often pressure the most when their children most need support, such as during sporting, cultural, and academic contests or performances.

Elizabeth, a thirty-one-year-old travel agent, is haunted by an event that happened when she was ten. Clad in black leotards for a gymnastics competition, she was standing on the balance beam when she heard her mother say to her father, almost in a stage whisper, "God, she's so awkward." Within moments, Elizabeth fell from the beam.

"I knew how not to fall off but I lost my concentration after hearing her," she says. "I have a picture of me that day. I looked so sad. After that, I didn't compete."

Just as Elizabeth suddenly lost her balance, children of Perfectionists often grow up feeling like physical, mental, or emotional klutzes. There are several reasons for this: They get little praise or constructive guidance from their self-involved parents; they are shamed for their failures, which makes it harder to try again; they live with anxiety, which makes it harder to feel at ease and perform well; and, for some children of Perfectionists, being less capable than their parents is a form of loyalty, allowing the parent to always be the one who looks graceful, brilliant, and in charge.

Elizabeth's Perfectionistic, Using mother demanded perfect obedience: "When my parents had company, my sister and I had to come curtsy good night to guests. We could help serve dinner but never be a part of it. It was like parading out pets. We were not valued for ourselves, but for being hers."

Her mother drilled Elizabeth in every social detail. After rehearsing with Elizabeth how to say hello when answering the phone, her mother would leave and call her daughter, hanging up and ringing back repeatedly until Elizabeth's hello was sufficiently "ladylike" and "friendly." Insisting that to be a good cook Elizabeth had to develop a keen sense of taste, her mother allowed her only one taste of each dinner item, after which Elizabeth had to name correctly every ingredient in the dish before she could go on eating.

Elizabeth's father practiced a different form of food control. He encouraged Elizabeth to cook desserts for him in home economics class, inviting her to eat the desserts with him. At the same time, he told her she was fat and that no boy would want to date her.

Elizabeth's father rarely seemed to have time for her, and when he did, he was impatient. Once, while her father was building a deck, nine-year-old Elizabeth asked if he would teach her how to pound nails so she could help him. When she tapped too gently on the first nail and it fell over, he grabbed the hammer from her and told her to leave him alone.

Pressure to Excel

Perfectionistic parents apply heavy pressure on their children when it comes to choices of schools, hobbies, and careers. When I was sixteen, on the morning of my first college interview, my father was up early, sternly finding fault with everything I did: how I tied my tie, how long I took in the shower, how quickly I ate. By the time I arrived at the interview, I was a wreck.

My father, a man of great wit, ambition, charm, and self-discipline, had always run our family much like his large corporation. At home, as at work, he valued winning, discipline, and the absolutism of his authority. From outward appearances, we were a successful, happy—even model—family. But inside, we were troubled. My father's love seemed of a strange sort, encrusted with tirades without warning, searing criticism, perfectionism, and ironclad rules. My mother, a sensitive soul, acquiesced to his domineering early on, and I was left confused about love, life, and myself. Like many controlled children, I second-guessed and self-blamed.

By the college interview, I'd had years to get used to my father's withering criticism. But you never get used to it; it always hurts. I was able to cope, even perform during the interview—children of controlling parents learn this well—and I was admitted to that college. That was all that mattered to my father. But to me, his behavior was cruel and confusing. I remember being puzzled, one of countless times I was puzzled in my youth. What had I done wrong? Why was he so upset? I was the one being interviewed, not him. More than anything, I wanted to hear my father say what children of Perfectionistic parents most yearn for but rarely hear: "Just do your best. I'll be proud of you no matter what."

Like many Perfectionistic parents, my father launched into tirades to discharge *his* anxiety about my performance. His behavior also reflected his lack of thought as to what I needed or wanted. Perfectionistic parents intermingle their own dreams with their children's lives and lack the perspective to acknowledge the difference.

Status Worship

Many Perfectionistic parents worship beauty, status, power, or money to the point where the coveted item takes on a near religious quality, revered beyond reason.

Brenda, now a fifty-four-year-old homemaker, is a frizzy redhead who grew up in Los Angeles, where her father had a small shop for which her mother kept the books. She recalls being aware even at age ten that her "Barbie-doll parents" were obsessed with the "look of Hollywood." But Brenda was born looking completely different from her second-generation Italian-immigrant parents: "I came out looking very Irish. Though some of the relatives lauded my red hair, pug nose, and freckled face, and endearingly call me 'Brick Top' or 'Rusty,' I felt ugly and worthless in my parents' eyes."

Brenda remembers having a hearty guffaw until her father said disgustedly, "Don't you have a different kind of laugh?"

One of the most costly traits of Perfectionistic families is that they dictate that not only their children's appearance and performance, but also their emotions, must be "perfect." "Imperfect" feelings like sadness, doubt, grief, anger, or fear are not tolerated. Perfectionistic households are grimly serious; Brenda cannot recall a time when her Perfectionistic, Depriving parents laughed. Today Brenda finds it hard to relax, laugh, or be spontaneous: "When someone takes away your laugh, they take away your soul."

Brenda's father frequently called her a "whore" and "damaged goods." At sixteen, as she waited outside a movie for her father to pick her up, a group of boys began sexually harassing her. She ran inside the lobby to escape, so her father drove past several times before Brenda saw him. He was furious at the delay and after she told him what had happened, he slapped her face and yelled, "You know why, don't you? You look like a little slut. I don't even want to look at you."

Disdain for "Flaws"

Sometimes a parent's silent disapproval of a child's "imperfections" can be as painful as scathing criticism.

Chip, twenty-nine, is finishing his junior college degree. His wealthy parents, mainstays of the Boston philanthropic scene, adopted him when they were physically unable to have children. But when Chip was diagnosed with learning and physical disabilities at age six, his father pulled away.

Each time Chip's accomplishments fell short of other children's, he felt like damaged goods. To this day, he wonders whether his parents' inability to have a child, compounded by adopting one who turned out to be "flawed," wounded his father's ego.

Because Chip felt as though his Perfectionistic father never believed in him, he has found it hard to believe in himself. At age twenty-nine, Chip is finishing a junior college degree. His ten-year path to get his degree has been studded with dozens of menial jobs, aimless travel, and drug use.

In recent months Chip has tried to make up for lost time. When he told his parents he hoped to go to a top-notch university and major in anthropology, both parents tried to dissuade him: "My dad told me, 'You're just setting yourself up for disappointment.'" But in our interview, Chip proudly told me of the letter of acceptance to the University of Michigan he'd received a few days earlier.

Obsessions

Many Perfectionistic parents seem obsessed with order and cleanliness.

Deirdre is a thirty-six-year-old office manager. When she was eleven, she and her brother went to live with her father and stepmother in Arkansas. Her stepmom immediately set down the rules. Clothes on hangers had to face the same way, buttons buttoned, zippers zipped. The pockets on Deirdre's clothes were sewn shut so they wouldn't get dirty or torn. Drawers were labeled, their contents organized by size. On the TV lay a pair of white gloves that the children had to use when they changed channels so they didn't get the TV dirty.

Deirdre recalls seeing her stepmom break down in tears when finding a single hair in a sink. Deirdre also remembers waking up at three A.M. more than once to the sound of her stepmother vacuuming, often following a marital spat.

Deirdre's father, a preacher, could not tolerate "downtime" among the children: "Dad would walk through the living room when we were watching TV and grumble, 'What are you going to do, watch that all day?' It got so when we heard him come home we'd turn off the TV and scramble to start housecleaning."

A certain amount of order is necessary in any home, but Perfectionistic parents, particularly those who also have Cultlike tendencies like Deirdre's stepmom's, have routines for eating, sleeping, cleaning, and talking that children disrupt at their peril. These parents seem panicked or enraged when something isn't where they expect it to be. They cannot bend with the innate disorderliness of children and life. By demanding excessive order, controlling parents act as if their children are furniture—which needs to be cleaned occasionally, but which doesn't make trouble and is always exactly where you left it. Controlling parents don't like their children to be too bubbly or rambunctious.

Deirdre still struggles with her own compulsive neatness: "Last week I found myself walking around after my boyfriend, straightening up each time he touched something."

Self-critical parents who see their children as extensions of themselves can't help but judge their children harshly. Perfectionistic parents may focus on children's failures and "flaws" as an excuse to withhold love. In so doing, they sidestep the vulnerabilities inherent in loving another. Yet because perfectionists find fault with everyone, they are always disappointed. They try to be perfect so others will love them, yet they can never succeed. They try to make their children perfect so they can love their children more, but both they and their children are destined to fail.

Self-Assessment

My parent(s):

- Pressured me to perform

- Demanded unrealistically high standards

- Did not tolerate flaws or mistakes

- Seemed obsessed with cleanliness and organization

- Seemed fixated with status, appearance, or prestige

Next: Cultlike Parenting

In their quest to escape flaws by demanding the best, Perfectionistic parents share similarities with the next style, Cultlike parents. Cultlike parents seek to escape uncertainty by always having to be "right" and "in the know."

5

CULTLIKE PARENTING
Obedience with a Missionary Zeal

The greater the ignorance the greater the dogmatism.
                    —SIR WILLIAM OSLER

Key Characteristics of Cultlike Parents:

- Control through ironclad adherence to rituals and beliefs
- Terrified of doubt and uncertainty
- Fear questions, dissent, or new ideas
- Seek security from organizations or philosophies

Potential Consequences of a Cultlike Upbringing:

- Reduced initiative
- Heightened distrust or gullibility
- Social isolation
- Distorted intellectual development
- Complicated spiritual life

Now a forty-four-year-old artist, Shirley remembers herself at seven, her honey-colored hair in prim pigtails, spending every day after school as a captive audience in the family garage. She dared not look bored, as her mother, a born-again Christian, lured neighborhood kids with lollipops so they would sit through her proselytizing at her "Good News Club"

*impromptu Bible classes. Eventually a school principal sent a flyer
warning parents to keep their children away from Shirley's mom: "My
mother had a way of frightening children with her constant praying,
threats of hellfire, and sudden outbursts of 'Jesus help me, I am a
worthless sinner!'"*

*Starting at age four, Shirley had to memorize a dozen Bible verses
weekly and attend church meetings four nights a week. Christmas
presents were forbidden after her mother realized that the word "Santa"
had the same letters as "Satan." Shirley and her brother could never eat
devil's food cake. Shirley remembers sometimes waking in the middle of
the night to find her mother "bending over me with a Bible, praying,
muttering and whispering to herself."*

Her mother carried to extremes a central trait of Cultlike parents:
the need to feel certain. Cultlike parents have to know "the truth" or
belong to a group that is "in the know." I use the term "Cultlike"
because these families mimic destructive cults in several ways:

1. Leaders act larger than life and receive special treatment.

2. Members' rights are subjugated for the "good" of the group or
   leader.

3. Prejudices rigidly separate members from outsiders.

4. Behavior is tightly regimented.

5. Feelings are devalued, minimized, and manipulated.

6. Questioning and dissent are not tolerated.

"My mother always told me that Jesus came first in her life,"
Shirley said. Yet her mother's devotion to Jesus left Shirley feeling con-
siderably less than a priority. As Shirley grew older, her mother viewed
her daughter's innocent questions about religion as blasphemy. Once,
while in her teens, Shirley brought home library books on witchcraft.
Her mother immediately summoned church friends for an emergency
exorcism: "I had to kneel for four hours. They were shouting, 'Do you
love Jesus?' 'Are you washed in the love of Jesus?' Finally I fainted
because of no food or water and they said, 'Praise God! The demons
have left her. This child has healed.'"

While her mother inundated Shirley with zealotry, her alcoholic,
Abusing father cowed Shirley with threats and violence. Once, during
a slumber party with two girlfriends, she and her guests were awak-

ened at two A.M. by shouting from her parents' bedroom: "My mother was quoting Bible verses and screaming, 'Satan, get thee behind me.' My father was throwing change at her, yelling, 'Do I have to pay for it?' Then he put his fist through the wall. I never asked friends over after that."

Today Shirley struggles with the lasting effects of her upbringing: depression, distrust of others, and sensitivity to criticism: "When people criticize or get angry with me, I melt down. Sometimes I'll believe them and buy whatever they say. Other times I'll just go off and cry.

"I feel like a concentration camp survivor," Shirley adds. "I was at my parents' mercy and they didn't have any."

Military Families

While some Cultlike parents use religion to know the "truth," others find certainty in institutions that allow them to know the "rules." The military, with its authoritarian structure and regulations, can attract Cultlike parents searching for a way in which to order their lives.

Caitlin, a forty-one-year-old teacher, recalls that as kids she and her six siblings were rousted out of bed on Saturdays at six A.M. by her father, a navy officer. The sleepy-eyed children were trotted to the kitchen and shown the "watch bill" of chores for the week. Military-style standards prevailed. Beds had to have hospital corners and bounce a dime. Garbage-can liners were to be folded and creased square at the corners. Silverware was organized in the dishwasher by implement. Milk was stored in a pitcher, never in the milk carton; butter on a plate, never in the carton.

Caitlin's Cultlike, Perfectionistic mother would answer her children only if addressed by the words, "Mom, may I speak?" Her father would ignore any statement lacking the prefaces "Father," "Dad," or "Sir."

Cultlike parents zealously adhere to rigid behaviors because they are troubled by the gray areas of existence. Knowing the rules allows them to view life in right-or-wrong, all-or-nothing terms. Furthermore, many controlling parents are not adept at communicating about personal issues and feelings. "In our family we never said, 'Let's talk things out,'" one woman told me. "'I don't understand' and 'Why?' were not a part of our vocabulary."

Among those I interviewed from Cultlike families, tension and the impending threat of physical or emotional violence pervaded their lives. "Our home was like the lid on a boiling pot," says Caitlin. But in true military fashion, the children were supposed to be stoic when punished: "You were the scourge of the earth if you cried. Crying was a sign of weakness."

Many controlling parents, not just Cultlike parents, interpret children's questions or lack of instant compliance as a deliberate challenge to their authority. They cannot see that in many cases their children are simply afraid, unsure, or preoccupied. When they feel deliberately challenged, some parents respond with violence.

This is not to say that children in controlling families don't seek out ways in which to keep alive the flame of their individuality. The day Caitlin graduated from high school she packed her suitcases. That summer, she slept next to her packed bags, counting the days until she could leave for college.

Her on-edge childhood has left Caitlin struggling with workaholic and perfectionistic tendencies. She has lived much of her life with a low-lying sense of fear and foreboding. Recently, trying to conquer her fears, she sought out the scariest challenge she could imagine and began taking sky-diving lessons. When interviewed, she had recently completed her first solo free-fall dive.

Fundamentalist Military Families

Jonathan, a thirty-five-year-old financial planner, grew up with a Cultlike double whammy: a zealous military father and a strict Catholic mother.

One bright Saturday morning when Jonathan was twelve, he dutifully spread newspapers on the kitchen floor and set out his father's scissors and razor. His father, an army officer, marched in and, as he did every other Saturday, cut Jonathan's hair in a half-inch butch cut.

The ritual devastated the boy, who hated his hair so short and, in addition, was forced to assemble the implements and clean up the results. As his father finished his haircut, his mother came in and asked Jonathan, "How does it look?" Near tears, Jonathan didn't answer. His father immediately shaved Jonathan's hair half again as short. "That's for not responding to your mother's question," he told his son.

Before dinner, Jonathan's father would shout, "Inspection: hands!" Jonathan and his brothers were to thrust their hands forward to show they'd been washed. "My father had a military model for how cadets were treated, and he applied it to us," Jonathan comments.

Jonathan's mother, a devout Catholic, blended military discipline with religion. On wash day his mother hovered as her children methodically folded linens three times, crisply, repeating "the Father," "the Son," and "the Holy Ghost" with each respective fold.

In junior high, Jonathan realized that he was gay. He felt that he could never tell his parents: "I was a people pleaser, always trying to smile, hungry for approval. It was hard for me to say what I thought." It wasn't until he was twenty-four that he came out to his parents. To his surprise, his father had little reaction, but his mother collapsed into the arms of a church friend and sobbed off and on for three days. Finally she said, "Jonathan, I know you *think* you're a homosexual . . ." Jonathan recollects that, "It went downhill from there. It epitomized her cookie-cutter mentality. Here was my mother thinking she knew more about her grown son's sexuality than he did."

Whenever Jonathan brought up his sexual orientation, his mother would quote from the Bible and try to talk him out of being gay. For a time, he stopped speaking with her. "That was hard because I knew family was so important to her," he admits. "But what hurt me more was that nothing I could do or say could change her mind. She was willing to lose her relationship with her son to keep her religious belief system."

After several months of little contact, relations slowly warmed as Jonathan and his mother agreed that discussions of his sexual orientation were off limits. Since then, he has spoken with his mother about how she failed to protect him from his father's abuse and how unhappy their "code of silence" about his being gay was making him. His mother apologized: "What she did say was from the heart. Now it's a relationship I can deal with. I used to cringe when I thought of talking with her."

As for his father, Jonathan struggles to make a connection. In a letter that he has yet to send, Jonathan writes, "Like a logger clear-cutting his way through a national treasure, you trampled me. That not being enough, it seems you've now discarded me."

Prominent Families

Some families prominent in social, political, or corporate circles also share a Cultlike style. Doing anything that would embarrass a prominent parent or hurt a parent's chances for corporate or political advancement is viewed as a mortal sin. These parents see their children's needs as secondary to the needs of the social circle or corporation. Children in these families can end up feeling like props.

Herb, forty-four and successful in the medical field, was a ten-year-old curly-haired boy with a cherubic expression when his Cultlike, Perfectionistic father stepped onto the corporate ladder of a large Midwestern manufacturing company. As his parents struggled to fit in with the corporate social set, Herb's and his younger brother's lives changed dramatically. Their father, consumed by his climb up the company hierarchy, intensely examined corporate nuances of office size, seating at meetings, and the makeup of golf foursomes to see who might be edging him out for advancement. This scrutiny eventually extended to Herb. He had to dress right—as well or better than other corporate sons—even down to his country-club swim trunks, which had to be ironed before every visit. He had to think right, getting only top-notch grades, since this might reflect on his father's chances for advancement. Most of all, he had to act right, behaving "like a perfect little gentleman." Before company social functions, Herb's father would rehearse how Herb should greet his father's superiors or anyone else his father wanted to impress, saying their names in a strong, clear voice and giving a firm handshake.

"Everyone talked about how well-behaved and good-looking my brother and I were and my parents just glowed," Herb says. "I felt like we were just dough to be molded into a final product."

Herb's father, like many Cultlike parents, needed to feel superior to others: "His attitude was, 'We are special, our race is special, our religion is special, and our corporation is special.' He would always remind me that we had an ancestor who was a signer of the Declaration of Independence. He'd tell me, 'Don't trust anyone whose name ends in a vowel.'"

Herb's father's infatuation with appearances left his son feeling anything but special: "I felt that I just didn't figure in his life. I don't know what it was about me that he objected to."

To compensate, Herb tried to be perfect. When he was a senior in high school, friends invited him to join the decades-old class ritual of painting their class year number on a local bridge. Herb asked his father for permission to go: "I felt like such a schlemiel. Here I was, asking permission to do something you're not even supposed to ask permission for and my father saying it was okay to do something like that once in a while."

Over time, Herb developed a "doofus" persona. He got depressed, became a loner, and abused drugs. Today, he still feels like a "black sheep" despite his postgraduate degree and good job.

Self-Assessment
My parent(s):

- Strongly identified with a military, social, religious, or corporate group or credo

- Tolerated little dissent, questioning, or uncertainty

- Distrusted strangers and "outsiders"

- Saw rules and beliefs as more important than relationships or feelings

- Viewed situations in black-and-white terms

Next: Chaotic Parenting
The next style of controlling parents, Chaotic parents, combines elements of two earlier styles, Smothering and Depriving parents. While Smothering parents overwhelm children with too much or the wrong kind of love and attention, and Depriving parents starve their children with too little love and attention, Chaotic parents both deprive and smother. Unable to maintain a constant demeanor, they seesaw between overwhelming closeness and rejecting withdrawal.

6

CHAOTIC PARENTING
Life in the Quicksand Lane

The weak can be terrible because they try furiously to appear strong.

—RABINDRANATH TAGORE

Key Characteristics of Chaotic Parents:

• Control through mystification and unpredictability

• Extreme difficulty setting consistent limits

• Use double binds, mixed messages, and bizarre reasoning

• Radically changeable behavior

Potential Consequences of a Chaotic Upbringing:

• Confusion about emotions

• Hypervigilance

• Reduced trust in others

• Life becomes an emotional roller coaster

Brittany, a twenty-three-year-old sales representative, can't put out of her mind the time she came home ten minutes past her curfew when she was sixteen. Her Chaotic, Abusing mom sprang from her chair and grabbed Brittany by the arm, wrenching and twisting, then digging her long fingernails into Brittany's flesh. Screaming that her daughter was a "tramp," she sent Brittany to her room. Forty-five minutes later her mother knocked on Brittany's door, and wearing a big smile carried in a gourmet dinner she'd cooked for her, complete with a rose in a bud vase. Brittany's childhood was full of such about-faces. It wasn't unusual for her alcoholic mother to ground Brittany, then within an hour tell her to go visit friends.

When Brittany was seventeen, her mother got her a prescription for birth control pills and proclaimed her progressiveness by letting Brittany's boyfriend spend the night with Brittany at their house. When the couple came down to breakfast the next morning, her mother called Brittany a "slutty whore" and threw the boyfriend out.

Brittany's mother bore a key trait of Chaotic parents: an inability to maintain an emotional middle ground. Unlike Smothering parents, who are too close, and Depriving parents, who are too distant, Chaotic parents encompass both extremes. "She'd go from babying me to practically kicking me out," Brittany exclaims.

Chaotic parents often mete out harsh punishment, then pander. In their effort to stay afloat in a churning emotional sea, Chaotic parents counterbalance their most recent emotional excess by racing to embrace its opposite. Out of desperation, they often solicit outside authorities for help. Brittany's mother once called the police after she had pulled Brittany's hair and her daughter had in turn scratched her. When her mother demanded that the police arrest Brittany, they refused. One of the officers took Brittany outside and gave her his card in case she ever again needed help with her mother.

When Brittany was three, a car accident put her in a coma. Doctors told her mother that Brittany had only a 25 percent chance of survival. Yet her mother insisted on every test possible and never left her daughter's bedside. Now Brittany wonders, "After such a close call, you'd think she'd be happy I lived. So why did she abuse me so?"

Brittany articulates the fundamental, terrifying question that haunts many controlled children:

If they love me, why do they hurt me?

The answers, too scary for young minds to contemplate and young hearts to accept, include:

Maybe they don't love me.

Maybe they don't want me.

Maybe they are not in control.

Maybe they hurt me for no reason.

Maybe they will hurt me no matter what I do.

Rather than face these thoughts, children adopt other answers, still discomforting but less terrifying:

They hurt me because I am bad.
They hurt me because I deserve it.
They hurt me because they love me.

Such answers allow children to grasp inexplicable events. If a child feels responsible for parental behavior, then he or she can control something by trying to change. Yet these rationalizations take a toll. Brittany still finds it extremely hard to tell when people are being duplicitous. She struggles to find solid emotional ground and suffers depression and low self-esteem.

Confusing Messages

Ina, a fifty-three-year-old social worker, was fourteen when she looked into her mother's taunting face, knowing chocolate cake was missing and that she, of all six children, was being called the thief. As Ina denied it and began to cry, her mother triumphantly said, "See, I knew you took it. If you weren't a thief, I wouldn't get a rise out of you."

Ina's childhood was saturated with no-win situations. Through bullying and crazy reasoning, Ina's mother put her children in constant binds. She warned Ina to watch out for "kidnappers," then gave her daughter adult-strength sleeping pills when Ina developed night terrors about being kidnapped. Her mother ordered Ina to be smart and pretty, with top grades and lots of dates, yet discouraged her from acting smart or feeling pretty around the house.

Curfews and other household rules were ignored one day, rigidly enforced the next: "I never knew what the rules were until I broke them. Sometimes I'd do something and get no response. The next time I'd do it, my mother would explode." If Ina disagreed with her mother, she was labeled a "paranoid schizophrenic"; her mother even invited neighbors over to watch her daughter "act crazy."

Ina walked on eggshells because she never knew the rules, just as Brittany never knew her mother's next mood. "For years I thought I was nearly crazy, but since my mother labeled me a paranoid schizophrenic I was ashamed to ask for help," Ina confesses. "It took years before I realized I was not crazy. I grew up in a crazy home. When she'd ridicule me, I'd tell her I didn't think she loved me. Then she'd say, 'How can you say that? I told the neighbors the other day how wonderful you are and how much I love you.' It left me feeling totally confused."

Chaotic parents like Ina's mother don't experience their inconsistent limits and mixed messages as erratic because their sense of who they are tends to vacillate. When they're sending a mixed message or enforcing a double standard, their actions are consistent with their momentary sense of self. As that sense of self changes, their actions change. Children of Chaotic parents can grow up thinking crazy things are normal because, for their parents, they were.

One major bind facing children of Chaotic parents is how to negotiate their needs for autonomy and intimacy. If a son or daughter wants to visit friends instead of staying home, Chaotic parents often act rejected and hurt. If the child stays home, Chaotic parents often become critical or rejecting. As a result, children of Chaotic parents feel guilty desiring either independence or closeness.

Chaotic Lifestyle

A similar sense of chaos can exist in some families with a parent who is mentally ill. While these parents are not to blame for their chaotic actions—they are ill, not bad—their controlling behavior can leave their children overwhelmed by chaos.

At ten, Celina, now a thirty-seven-year-old teacher, stood at the edge of the living room gathering up the courage to ask her mother a question, knowing requests often sent her mother into a panic. The family apartment was steadily accumulating stacks of flyers and newspapers and degenerating into chaos, and Celina needed a spot where she could practice her clarinet. Finally, she meekly asked for practice space. This time, unaccountably, her mother said yes. Many other times she had said no.

Celina's mother was a paranoid schizophrenic who had emerged, ostensibly stable, from a psychiatric hospital two years earlier. Estranged from Celina's father, she raised her daughters on her own, and although earlier considered mentally stable, signs of her mental illness soon reappeared. She neatly stacked and labeled Celina's toys, but didn't allow Celina to play with them. She would buy three sizes of each item in Celina's wardrobe because she was paranoid about Celina's trying on clothes in the store. Celina and her brother couldn't go in certain rooms of the house because their mother feared that exotic diseases lurked there. The family often dined out because their mother was afraid to eat at home. Celina spent much of her childhood doing homework in restaurants.

*Summertime threw Celina's mother into a panic over planning what
to do with the kids when school was out. Fall upset her because she had
to go to the store to get new school clothes. When Celina was thirteen and
asked for her first bra, her mother flew into a rage. "Bras? We can't have
bras in this house," she screamed. Celina only got a bra when it was
required for gym class.*

The chaos took an early toll. Celina recalls being four years old,
playing on the floor, when her mother and father began a vicious
screaming match: "I remember taking some part of myself and burying
it in the wood floor, pretending I was an innocent child and didn't
notice the chaos. It's as if a part of me is still buried away in the floor."
Celina's mother was desperately trying to control a world in
which, because of her mental illness, she felt in free fall. "It was scary
that she could change from an incredible yumminess to being a terrifying creature," Celina recalls.
Her mother also stressed that Celina not draw attention to herself,
which further confused Celina. Once, after enrolling Celina in an acting class, she warned her daughter to "be inconspicuous" in the class. At
times the older woman seemed to think her children were parts of her.
When people asked Celina how she was feeling, her mother would say,
"We're tired." As a result, Celina had difficulty in individuating. As a
grade-schooler answering a question in class, Celina would begin, "My
mother says . . ."
When Celina's mother's condition worsened, she threatened to
turn on the gas and kill herself and her children if Celina's father
returned and tried to take them away. Petrified, Celina would wait
until her mother was asleep, then tiptoe to a window and open it. She
did this even during frigid Minneapolis winters in case her mother
decided that that night was the night to turn on the gas. Still, Celina
didn't tell anyone about her chaotic home life for fear that she would
be put in foster care.
Today Celina is trying to heal through therapy, support groups, and
friendships. "I've always felt as if I have a Swiss bank account about
learning, loving, and intimacy but nobody has given me the account
number," she shares. "I want the number that will open my Swiss bank
account."
Celina's case is, of course, an extreme example, since schizophre-
nia is at one extreme of the mental illness spectrum. Yet the common
thread among Chaotic parents is communication rife with double
binds and mixed messages. Chaotic parents are bandied about by feel-

ings, fears, and needs, and that mercurialness envelops their children in a cyclone of confusion.

Self-Assessment
My parent(s):

- Had trouble finding an emotional middle ground
- Put out confusing and erratic rules and messages
- Applied discipline in ways that were too harsh, too loose, or inconsistent
- Showed abrupt about-faces in behavior and feelings
- Alternated between blind acceptance and total rejection of me and others

Next: Using Parenting
The next group of controlling parents, Using parents, actively take from their children in order to satisfy their own needs.

7

USING PARENTING
"Me First" Child Raising

The greater the power, the more dangerous the abuse.

—EDMUND BURKE

Key Characteristics of Using Parents:

- Control through demands for loyalty, admiration, and obedience
- Terrified of losing or feeling one down
- Emotionally immature
- Insensitive to others' needs and feelings

Potential Consequences of a Using Upbringing:

- Feeling used
- Poor self-image
- Mistaken ideas about love
- Difficulty in developing good emotional self-care habits

When Ellen, a forty-nine-year-old volunteer worker, was nine, she stood at the vanity mirror putting curlers in her mother's hair. Next, she plucked her mother's eyebrows and painted her nails. Later, Ellen sat next to the tub while her mother took a bath and talked about life. Finally, she brushed her mother's hair and told her how beautiful she looked.

For another mother and daughter, such a scene might bring the coziness of mutual sharing and intimacy; for Ellen there was nothing cozy or reciprocal about the evening ritual. If she put a curler in wrong she'd get a tirade or a slap in the face. Other than her scripted comments about how beautiful her mother looked, Ellen was expected to be silent while her mother complained about real or imagined slights.

Using parents tend to see life in terms of what they can get out of it. When children's needs conflict with parental needs, the children are seen as nuisances, problems, or threats. Like Ellen, children of Using parents spend their lives feeding their parents' hunger for attention, approval, and love.

Ellen's Using, Abusing mother used her daughter as servant, listener, and emotional punching bag. She called Ellen a "mistake" and often told her, "I gave up a good life for you." She frequently blamed Ellen for the C-section scar she received in giving birth to her. This is a classic Using-parent bind: Ellen was made to feel that her very existence had hurt her mother, an act over which Ellen had no control and one that she could do nothing to remedy.

When Ellen was eight, her mother, embarrassed and furious because Ellen swam poorly in a swim meet, dragged her home and forced her to take off her swimsuit, walk naked to the garage, and throw the suit in the trash. She often punished her daughter by backing her in a corner and forcing her to drop her hands so that she could slap her. The slaps hurt, but making Ellen drop her hands added a deeper injury because it blocked her instinct for self-defense. By teaching children not to defend themselves, Using parents increase the risk that their children may end up in unhealthy relationships or may develop self-abusive behaviors.

Ellen's father had died when she was two and her mother had quickly remarried. Claiming Ellen's dad had been unfaithful, her mother destroyed all pictures of him. As Ellen grew up, she hungered to know more about her father, but her mother would never discuss him. After an aunt told Ellen she looked like her father, she asked her mother, begging, "Do I look like Dad?"

To which her mother coolly replied, "I don't know. I never noticed."

Using parents' self-esteem rises a notch when they put someone else down. Even today, when Ellen wears a dress her mother likes, the older woman angrily demands, "Why didn't you buy me that?" When Ellen wears a dress her mother dislikes, her mother snidely asks, "What's that you're wearing? You look like a clown."

When Using parents feel slighted, they often become enraged. Many a child of Using parents can tell stories of going into restaurants or stores and watching, petrified, as their parents fumed over perceived bad service, bullying waiters or clerks whom they saw as too slow or not sufficiently deferential.

Using parents themselves were generally terribly misused or tormented as children. Because they received little approval, they grew up feeling unworthy and inadequate. As a result, they demand worship from their children—a reflection of their desperate need for self-esteem. Using parents are justifiably angry about their pasts, yet despite their great capacity for gaining access to their rage, they seem remarkably disconnected from most emotions. Ellen recalled, "Sometimes my mother would look really mad and I'd ask her if she was upset. But she'd always deny it, so I could never figure out if my perceptions were true or false." To Ellen, her mother seemed bitter and self-centered: "She seemed to have no reference for how she treated people and how others saw her."

Her mother's behavior cost Ellen dearly. In school she was the model student—quiet, obedient, and with good grades—but had no friends: "I never looked up. I had zero personality." She was afraid to dance, date, or listen to rock and roll. Her childhood was riddled with allergies, sleep disorders, chronic headaches, and eczema, all of which cleared up when she went away to college.

Self-Centered

Magda, a thirty-six-year-old civil servant, recalls her sixth birthday as an occasion for joy—for her father. He bought her a box kite, something he'd always wanted for himself. "He always got me things he wanted. He would play with them until they broke, all the while ignoring me and any questions I had. Afterward, he'd think we'd had a good time."

At first, Magda's father had a soft spot for his daughter. "He was quite taken with me as a baby," she says today. "He'd get down on the floor and do baby talk with me." But as Magda grew older and began to be more independent, her dad seemed to lose interest: "I was a cute, adorable little pet until I started developing my own opinions." When Magda was nine, her father left the family.

Magda's father had a classic trait of Using parents: immaturity. By ostensibly choosing toys for her that he played with himself, he reverted to being a child. Using parents give themselves free rein to have tantrums, expecting others to compensate for their excesses. Children of Using parents rarely have a chance to be children because Using parents take up all the room for childishness in the family.

Like Magda's father, many Using parents see their newborn children as "blank canvases" who are totally dependent on them and on whom they can make their mark. Many derive a sense of mission from the early days of parenting; they see it as a "project" they imagine will fill their unmet needs or distract them from their problems. But when the children begin developing stronger identities and move toward independence, their parents feel angry and betrayed. Feeling rejected by their "creations," these parents emotionally—and sometimes physically—abandon them.

Magda's adult legacy has been relationships with men in which she has felt little right to voice her needs—and often fears being deserted by partners after even a minor disagreement.

Jealous

Using parents see life as a series of situations in which only one person can win and victory must come at others' expense. The parents don't want to lose, even if it means their children end up the losers. This may explain why such parents often seem jealous of others' success or good fortune and become intent on spoiling others' joy.

On her high school graduation night, Robin, now a fifty-three-year-old design artist, proudly clutched her red-and-white tassel as she climbed into the backseat of her parents' car. After the ceremony, her mother grumbled to her father, "Well, I suppose we have to do something special for her." As they drove around aimlessly, Robin's parents began arguing over what they should do to celebrate. Each parent vetoed the other's suggestions, with her mother and father eventually screaming at each other. Robin cowered in the backseat, tearfully asking to be taken home.

The evening had been only the latest in a long string of burst bubbles. When Robin brought home an eighth-grade report card with a teacher's note reading, "You deserve all the luck in the world," her mother snapped, "Luck doesn't just happen; you have to go out and make it happen."

Often it seemed that her mother was downright mean. When Robin was twelve, her piano teacher told her she had real musical promise. Within a week, her mother, who seemed jealous anytime her daughter was acknowledged by others, stopped paying for Robin's lessons.

Her mother seemed jealous of Robin's relationship with her stockbroker father. When Robin would sit raptly listening to him talk about his day, her mother would shoot her dagger looks and would be cool toward her for days after. And, when Robin got engaged, her mother's first words were, "What if you discover something bad about him?"

Robin grew up feeling terribly alone. During our interview, she tearfully told me that when she was ten, she'd take a floor mop out of the closet for solace: "I'd pretend the mop was a twin sister and my best friend. Can you imagine—a floor mop?"

Robin's Using, Depriving mother was also hypersensitive to real or imagined slights from others: "After interactions with people, my mother would say, 'Did you see how she looked at me? Did you see how she talked to me? Did you see how anxious they were to leave?'"

Robin dutifully listened to her mother but had nobody to listen to her: "No one noticed my pain. I tried so hard to be good. I always blamed myself for not being perfect and lovable." Her self-image was so poor that when a junior high counselor announced she was going to be part of a special class, Robin assumed it was for "dumb kids." After the counselor told her that it was for smarter children, she replied, "No, that can't be. I'm too dumb."

Recalls Robin: "I was dumbfounded. I had no clue as to my abilities."

Using parents tend to see life primarily in terms of how it affects them. At a time when Robin was nervously caring for her six-week-old infant, suffering from the flu, and facing a stack of unpaid bills, her mother visited, expecting to be waited on. Seeing her mother giddily playing with the baby, Robin asked, "How come you never treated me like that?" Her mother departed, screaming, and wouldn't talk to Robin for weeks.

Using parents have little ability to see their children's emotional needs because they cannot consistently provide for themselves emotionally. If they were wounded as children and never got help, their wounds developed into an emotional abyss of unfulfilled dreams and unmet needs. Using parents are terrified of exploring this void; instead, they look to others to fill it. Children of Using parents learn that their primary job is to do nothing that will pose a problem for their parents.

Perhaps because Using parents feel empty, they try to accumulate wealth and status—just as some Perfectionistic and Cultlike parents do. They tend to have contempt for others who do or have less than they do, envy others who do or have more, and possess an underlying fear that if they were to lose their own status or belongings they'd be worthless. Robin's mother, for example, talked longingly of her youth, telling stories of servants, beautiful linens, and being treated like a princess. Yet Robin wonders how much of this was true; an aunt once hinted that Robin's mother grew up with cold, remote parents and was often shuttled off to relatives for months at a time without any explanation.

Just as perplexing to their children, many Using parents often are admired by others from outside the family. The children wonder what's wrong with them because they see a tormentor instead of the charming, witty, attractive mother or father others tell them they are so lucky to have. As we'll see in Part Three, healing from a controlling parent involves, in part, honoring your own experience and perceptions, despite others' views, so that you can have a full-palette view of your parents.

Narcissistic

Although the style of Using parents is most similar to what we think of as unhealthy narcissism, all eight styles of controlling parents share it. Narcissism is a distorted sense of self that leads one to see and treat others as unequal. The American Psychiatric Association's fourth *Diagnostic and Statistical Manual of Mental Disorders* (661) describes narcissistic individuals as those who tend to:

- Have a grandiose sense of self-importance

- Be preoccupied with fantasies of unlimited power or beauty

- Believe they are special and unique

- Require excessive admiration

- Feel entitled to be obeyed

- Take advantage of others

- Be unwilling or unable to recognize others' needs and feelings

- Envy others or believe others envy them

- Act arrogantly

Psychiatrist Charles Whitfield, in *Boundaries and Relationships* (134), observed that narcissistic parents tend to:

- See others primarily in terms of whether they can be useful or threatening
- Blame others rather than accept responsibility
- Try to dominate and control people, places, and things
- Be emotionally unaware, numb, or hypersensitive
- Project their character defects onto others
- Be hypersensitive to criticism or rejection
- Inappropriately express or internalize their anger
- Be perfectionistic
- Lack empathy
- Feel draining to be around

One of the most poignant moments in all my interviews was at the end of my interview with Ellen, whose mother, you'll recall, was a notably narcissistic parent who blamed her daughter for the cesarean she was forced to have in giving birth to her.

Ellen, twenty-nine years after dropping out, has returned to college to get a degree in art, her lifelong love. As we talked, she sat on her apartment patio by the water, the setting sun painting her profile a rosy orange. During our interview Ellen's forty-nine-year-old face at times looked sixty-nine, creased by exhaustion and grief. At other times she looked twenty-nine, peaceful and calm as she spoke of the strength and peace she has derived from setting emotional boundaries between herself and her mother.

"I used to try and recall happy memories with my mother but I gave up. It hurt too much that I had a mom who never once said, 'How are you today? You look wonderful!'" Ellen said. "I have never confronted her with all her abuse. She would probably not even know what I was talking about. So, limited contact is probably the best solution for me."

When her mother dies, Ellen expects she will have much sadness but few regrets about having sought an arm's-length stance from her mother: "Maybe in my memories she'll be kinder than she has been in reality."

I asked Ellen if there were anything else she wanted to add. Meditatively running her finger along the rim of her teacup, she took a long breath and confessed, "I really, really loved my mother. But my mother didn't take care of that love for so many years. Eventually, my love just went away."

Self-Assessment

My parent(s):

- Demanded loyalty, attention, and admiration
- Competed with me or tried to spoil my happy occasions
- Used other people to satisfy their own needs
- Seemed hypercritical or hypersensitive
- Seemed immature, self-centered, or childish

Next: Abusing Parenting

The next style, Abusing parents, embodies many characteristics of Using parents. But while Using parents do most of their damage through emotional abuse, Abusing parents add physical or sexual abuse or intimidation to the equation.

8

ABUSING PARENTING
"Do It or Else" Child Raising

I would have been better off raised by wolves.

—ROSEMARY, 55, A MANAGER

Key Characteristics of Abusing Parents:

- Control through brute force
- Blame their children for "making" them abuse
- Feel they have the right to abuse
- Have poor impulse control

Potential Consequences of an Abusive Upbringing:

- Depression
- Addictions
- Hypervigilance
- Assumption that abuse is deserved
- Difficulty in trusting others

Jorge, now a thirty-two-year-old psychiatric aide, was twelve and had just come home from school when his mother shrieked his name from her bedroom. That was the signal for him to come and be hit. As always, he resolved not to cry when his mother hurt him.

On this day some of his mother's chocolates were missing. As Jorge slowly drew alongside his mother's bed, she dug her nails into his flesh, then twisted his arm. But that wasn't enough. Today would be worse than usual. She grabbed him by the ear, then held his hand over a lit kitchen gas burner.

Pulling him back to her bedroom, she got out a belt, ordered him to stand spread-eagled in the doorway, and began whipping him across his back.

Finally, he cried.

Eventually, she stopped.

His mother ordered Jorge to stand where he was until she told him he could move. Two hours later, when his father came in from work, Jorge was still standing in the doorway. His father walked past him wordlessly, as if Jorge were a mere ghost.

That night was one of many Jorge fell asleep clutching a picture of Jesus. Jorge thought he was being hurt by his parents because he was a bad person. He prayed that some of Jesus's goodness would rub off.

Jorge was repeatedly abused by his Abusing, Chaotic mother; his father did nothing about it. Sometimes, when he was alone with his father, Jorge would ask why his mother hurt him and why his father didn't stop it. "That's just the way she is," his father told him. "Behave yourself and let her do it."

Abusing parents do things to children that are hard to believe. After thousands of hours of therapy, I still sometimes find incredible some of the horror stories clients tell me—and, if anything, the horror and pain are often downplayed or minimized. Despite the increasing attention paid to child abuse, *tyrantosaurus* parents like Jorge's are far from extinct. In 1997, an estimated 1 million children in the United States suffered from child abuse in which excess control was a key factor.

One thing is a constant in Abusing families: Abused children grow up thinking they deserve the abuse. Jorge, who now works with autistic children, told me, "I used to think that if I could just go over a mistake fifty times and think why I did it, I could get rid of my imperfections."

As an adult Jorge struggles with depression, though he is in therapy and has confronted his mother and father about their abuse. He worries that the pain is so deep he may never recover: "Maybe it's too late for me. Maybe I should concentrate on my children. Maybe only they can make their way out of this family legacy."

Destructive

What sets Abusing parents apart from other styles of controlling parents is their destructiveness, for they seem bent on destroying their children. While Using parents exploit their children's innate loyalty to get them to go along with being used, Abusing parents simply overpower their sons and daughters.

At fourteen, Eve, now a forty-four-year-old secretary, sat in her best red dress at the dinner table Christmas Eve as her father carved the turkey for more than a dozen relatives, including three young nieces and nephews. Earlier that week her father had ordered Eve to give away her pet hamster, but she hadn't had the heart to do it.

As he cut into the turkey, he looked directly at Eve. Holding up a turkey leg with the carving knife, he shouted, "You didn't get rid of the hamster, like I told you. I'm going to cut it up with this knife. I'm going to wring its neck." Eve remembers her cousins squirming and beginning to cry while the other adults looked on in stunned silence. Nobody said anything to her father.

Caring for small, fuzzy things helped Eve survive emotionally. She put milk out for neighborhood cats; at one point, by her count, she had seventeen cats coming for dinner. But her father threatened to poison or shoot them. Once he caught a stray dog, tied it to a pole, and "beat it to insanity," Eve recalled. To protect the cats, Eve stopped putting out milk and frantically shooed them away when they kept coming.

As a girl, Eve lived in such fear that she spent many waking moments outside her home, hiding in the bushes and pretending they were her real home: "I had one bush be the kitchen and another be the bedroom. I was much more comfortable outside than inside."

When Eve was nine, her father threw her across the room and dislocated her hip because he didn't like the way she said "Yes, sir" to him. The next day, he told her she had to quit gymnastics before she got injured again: "I think he really believed that I dislocated my hip in gymnastics. I don't think he even remembered doing it to me."

Her father's control extended to her social life. When Eve dated a "hippie" against his orders, her father hit her. When she dated an African American, he attacked her with a steel pipe. When she married a Latino, he threatened to take her out of his will: "I don't think he did it, though. He wouldn't have wanted to spend the eighty dollars to have his lawyer change it. But he resented any freedom I had."

Eve's father was as destructive as any parent of the people I interviewed. His abuse induced a legacy of low self-esteem that keeps Eve working in clerical jobs even after getting a master's degree. Her upbringing has cost Eve mightily in her relationships with men. She has a permanent restraining order against one former lover who beat her. Prior to that, she was married to an alcoholic who battered her for years until she left him. He later killed himself.

While Eve still fears her father, she turned the tables a few years ago by fighting back. One day her father began chasing her with a lead pipe, threatening to kill her. "I grabbed his arm, stopped his swinging, and gouged him with my nails," she announces proudly. "He took photos, told all the neighbors I had beaten him up, and said he was going to build a court case. But the balance had shifted. I was no longer going to give in." Since Eve fought back, there has been no more physical violence between them.

Taking

Some Abusing parents seem bent on taking from their children by every possible method. They are like Using parents, but in the extreme.

Rosemary, now a fifty-five-year-old manager, recalls sitting at the dinner table at the age of nine, watching her six-year-old sister, who was beginning to look ill. Her mother was talking incessantly, unaware of her daughters. Next to her mother's plate was a razor strop that she brought to every meal so she could whip the girls if they spilled anything or didn't finish their food.

Her younger sister, Rosemary would realize years later, was developing an eating disorder from the tension at the table. She would gag on her food, try to swallow it for fear of getting beaten, then gag again. When her mother wasn't watching, Rosemary would sneak some of her sister's food and eat it for her. Their father noticed but said nothing.

Over the years, Rosemary became obese. Her sister became anorectic. Many nights Rosemary's sister would throw up in bed. Her mother would come in and beat the younger girl for throwing up. Eventually the sisters took a pail to their bedroom so their mother wouldn't know when Rosemary's sister vomited.

Rosemary's maltreatment was pervasive. Her Abusing, Using mother would punch her, then make her rehearse the story she'd tell outsiders—that she'd walked into a door: "If I ever said anything back, she'd crack me across the face. I'd have a purple face and people would stare at me like I was the Elephant Man."

Once when Rosemary was singing the happy-go-lucky rock song "Personality" while doing housework, her mother slapped her several times, and shouted, "I'll give you personality!"

"She couldn't stand to see me happy," Rosemary says sadly.

Another time, her mother ripped a red bow out of Rosemary's hair, screaming, "You tramp. You whore. You know what this means to men? You are filthy and disgusting." Yet her mother did nothing to stop a sixteen-year-old cousin from molesting eight-year-old Rosemary during their "naps" together.

For most of her life Rosemary has suffered from depression, eating disorders, obesity, alcohol dependence, loneliness, and thoughts of suicide. Some days she wishes her mother would die. Other days, she feels guilt-ridden for having such thoughts.

Several years of therapy have brought Rosemary to the point of being, as she puts it, "semi-human," but she cannot make up for her lost years: "It was like being in prison. She was the warden and she had no intention of letting me out. I would have been better off raised by wolves."

Poor Impulse Control

Patty, who is now a fifty-three-year-old counselor, was in seventh grade when she walked into the living room after school; her dad, slumped in his easy chair in front of the television, asked her the time. "Three-thirty," she told him, since she had glanced at the kitchen clock a moment before.

"I asked you what time it is!" her father screamed. "Don't make things up. You go look at a clock."

Patty tried to tell him she'd just checked the clock, but her father kept screaming until she went back, looked again, and told him it was now 3:31.

Recalls Patty, "I guess my behavior looked like disobedience to him."

The rage reactions of Abusing parents can be so sudden and inexplicable that we can only speculate on what triggers them. Patty's father may have expected to see his daughter reverse course and look at a clock. Perhaps that would have comforted him because he would have felt he controlled her actions as he would a servant's. When Patty did something unexpected, like answering without looking, it may have taken him by surprise, something most controlling parents do not like, since it upsets their sense of being in control.

Patty's Abusing, Depriving father had several methods of intimidation. He would grab his daughter and hold a lit cigarette an inch from her arm, then say, "If you move your arm, it's your fault because you're burning yourself." He tried the tactic on Patty's cousin once and burned the child when he jerked away. When Patty was six, her father put her on his bike handlebars and rode downhill at full speed. During those times plummeting downhill or staring at a hot cigarette, Patty felt completely in her father's control. She dissociated: "I floated up around the ceiling somewhere. I didn't identify with the person who was me. I'd look down and think, 'Look at that pitiful person crying.'"

Her father also ridiculed emotions. When five-year-old Patty became afraid of spiders, her father commanded her not to be afraid. When a sixth-grade teacher recommended that Patty see a counselor because she was so shy, her father commanded her not to be shy. "For my father," she now realizes, "emotions were things you could just command."

As an adult, expressing her feelings is one of Patty's biggest struggles. A women's group and twelve-step programs have helped her to feel whole: "In my women's group I realized I could cry and people would not reject me. If anything, they drew closer. It was a shock, because as a child my crying made my mother withdraw and my father furious."

Physical and sexual violence is often the only close contact abused children can get with a parent. "My dad's physical roughhousing was the only time he paid attention to me, so sometimes I'd even start it," Patty admits. "I'd end up getting bruised and hurt but I desperately wanted some closeness. Afterward my mother would say, 'Oh, you bruise so easily.'"

Why Parents Abuse

Abusing parents maltreat their children because they can get away with it; they're bigger and have the power. Many abusers cannot maintain a consistent sense of themselves or others. At times they realize that their children are delicate, dependent beings, but when the abuse impulse gets triggered, they see their children's innocent behaviors as deliberate provocations. In those moments, they no longer see their children as preoccupied, forgetful, and dependent creatures who want parental approval. Abusing parents do not regard their children as human.

Abusers also tend to experience guilt differently from healthier parents. When a healthier parent hurts a child, he or she will generally be troubled by the action and try to atone. But Abusing parents justify their actions based on what the child did "wrong."

Abuse runs across a continuum, and most of the people I interviewed experienced less physical violence than Jorge, Eve, Rosemary, and Patty. Yet they hurt just as much and struggle with just as many limits in their lives.

Clients who come from controlling families without physical violence often tell me, "It wasn't so bad. Nobody ever hit me or molested me." But they wonder why they suffer from fallout as lasting as those who were physically or sexually abused. In my experience, wounds from emotional abuse and control can last long and cut deep. Jorge, Eve, Rosemary, and Patty each admitted that their burns and bruises hurt less than the pain of feeling abandoned, degraded, and betrayed.

We tend to discount the power of verbal abuse and emotional tyranny, perhaps because of the absence of visible bruises. We know sticks and stones break bones, but we forget that names really can hurt us. Labeling a child—"lazy," "spoiled," "stupid," "ugly," "bad," "dummy," "crazy," "whore," "selfish," and a "mistake" topped the "hit" parade among those I interviewed—shatters self-image.

Threats of violence force children to watch their step and feel under scrutiny even if a parent isn't around. While the sting or bruise from a blow eventually fades, the threat of violence does not. This can translate into a generalized fear that the world is not safe.

The most profound violence of abuse is the underlying message it sends to children:

The person who is supposed to protect me hurts me, and there's nothing I can do about it.

Self-Assessment
My parent(s):

- Physically or sexually hurt or bullied me
- Insulted me or called me horrible names
- Severely overreacted to dissent or disobedience
- Lost control at the drop of a hat

- Seemed oblivious to or showed little remorse over the effects of their abuse

Next: Childlike Parenting

Abusing parents like Jorge's mom or Patty's dad often marry someone who does little to stop the abuse. Like alcoholics who find a codependent to subtly allow them to abuse liquor, Abusing parents, in part, are able to continue mistreatment because their partners allow it. Parents who stand idly by while their children are hurt characterize the final type of controlling parents, Childlike parents. Childlike parents seem fragile, mere shadows of human beings.

9

CHILDLIKE PARENTING
"Can't Do" Child Raising

Life really terrified my mother.

—EVELYN, 46, A NURSE

Key Characteristics of Childlike Parents:

- Control through inducing guilt in others
- Seem incapable of being adults or parents
- Receive caretaking instead of provide it
- Often seek Perfectionistic, Cultlike, Using, or Abusing spouses

Potential Consequences of a Childlike Upbringing:

- Few opportunities to be a carefree child
- A tendency to put others first
- Difficulty in expressing anger or resentment

When she was six, Evelyn, now a forty-six-year-old nurse, stood in line with her brother, mom, and dad at Disneyland's elevated Skyway ride. She could feel her mother's long skirt billowing against her back in the hot summer breeze as they neared the front of the line. When her father hopped in the first available car, Evelyn's mother shoved her in with him. The door clanged shut as Evelyn watched her mother and brother scramble into the next car. For the entire ride, Evelyn's father, who loved to taunt and scare others, swung the caged car back and forth, ignoring

Evelyn's tears and shrieks. Years later, she was able to understand what
had happened: Her mother had sacrificed Evelyn because her mother
was afraid to be in the cage with him.

"Life really terrified my mother," Evelyn recalls. Her mother, who took
Valium for years, would walk the three miles to town rather than take a
bus because she didn't know what to say if the bus driver said "Good
morning." Once, when eight-year-old Evelyn wanted to play with her
brother rather than accompany her mother to the store, her mother wept
for three days.

Evelyn's mother excused her father's abuse of Evelyn by saying, "He
can't help it. He grew up with a drunk for a dad." At ten, Evelyn wanted
to write to Ann Landers for advice on how to cope with her father. Since
her father read all mail coming to and going from the house, Evelyn
asked her mother's help in mailing the letter. Her mother refused, telling
her daughter to "pray to God."

When Evelyn reached puberty, her father would come up and lift her
blouse and make her stand in the living room while he ridiculed her
"small tits." When this happened, Evelyn's mother took two actions.

First, she closed the living room blinds so the neighbors would not see.
Then she left the room.

Unfortunately, Evelyn's mother, Loretta, was not qualified to protect and nurture her children. From her daughter's description, it
appears that Loretta was depressed and anxious for most of her life. In
part, this may have resulted from Loretta's imprisonment from ages
two to four in a World War II Japanese concentration camp, days away
from death from malnutrition when she was finally freed. It's sad,
because Loretta's moods and phobias could probably be successfully
treated today with a combination of psychotherapy and medication.

Because Childlike parents are scared and needy, they often play a
childlike role, leaving the caretaking to their children, a role reversal that
robs children of their youth. While Childlike parents may not seem controlling, they play a crucial role in unhealthy family dynamics. For example, they often gravitate to a strong-willed spouse—frequently a
Cultlike, Perfectionistic, Using, or Abusing personality. Drawn to the
apparent strength and certainty of the stronger spouse, they feel secure
with someone who will run interference, be certain when they are
unsure, and act big when they feel small.

Like Evelyn, children of marriages between Abusing and Childlike
parents are deeply deprived, assuming the blame for their abuse since
they get no positive messages from either parent.

Rather than intervening when a spouse victimizes a daughter or son, Childlike parents forfeit their children to the spouse's control. On top of it all, many such parents demand sympathy for their own problems. "I'm sick," they say. "I'm afraid." "I'm lonely." "I'm depressed."

Evelyn's caretaking of her mother may have influenced her choice of nursing as a career. "I certainly had lots of experience taking care of others," she declared. While we can have compassion for Evelyn's mother's depression and fears, we must regret the price Evelyn paid for her mom's limitations. For much of her life, Evelyn has tended to put others' needs before her own and has struggled with career burnout and one-sided relationships.

Terrified and Anxious

Molly is a thirty-three-year-old vegetable grower who recalls a recent visit during which her mother rocked back and forth with intense anxiety. "I know there's something I'm supposed to be doing," her mother, Lucy, insisted. "I can't just sit here." Lucy was terrified of her Cultlike, Abusing husband and lived like a slave. He monitored Lucy's phone calls, which she could only make and receive during specified hours. He dictated her errands and declared exactly how long it would take her to do them. He checked on her by phone several times a day.

Molly recalls seeing stark terror on her mother's face if a spot of grease splattered on the stove in the middle of cooking. If her father saw one grease spot, he'd launch into a tirade.

When she was in her twenties, Molly brought her mother the book Men Who Hate Women & the Women Who Love Them, *hoping her mother would get some insight into her marital relationship. Lucy's response was to beg Molly to take the book away before her husband found it.*

After Molly was born, Lucy suffered severe postpartum depression, sobbing constantly. She was also so sensitive that if Molly or her siblings said they didn't like her cooking, she'd burst into tears. Since Lucy couldn't deal with disciplining Molly or her brother, and her husband relished inflicting punishment, she turned the disciplining over to him.

Terrified of strong emotions, especially anger, sadness, fear, and joy, Lucy discouraged Molly from expressing feelings loudly or strongly. Molly never knew whether it was because strong emotions scared her or because Lucy was afraid they would trigger Molly's father's explosive temper. Perhaps both.

Lucy carried her self-effacing ways to her deathbed—succumbing alone in a hospital at age seventy-two after sending Molly away because she didn't "want to be a burden" to her daughter.

Fail to Protect Their Children

By no means is every husband or wife of an abusive parent a Childlike parent. Healthier parents stand up to an abusive spouse and succeed in deflecting the maltreatment, often at their own expense when the abuser turns on them. Others take their children and leave. Some spouses of abusers, while feeling powerless to leave or stop the abuse, at least tell the children that the abuse is wrong. It can make a world of difference to abused children if a parent tells them that they are not bad, that they do not deserve to be hurt, and that the abusive parent is the one with the problems.

Childlike parents, by contrast, rarely stand up to the abuser or try to undo it. Some even agree with the spouse's abusive punishment but don't want to have to deliver it themselves. Others make excuses for the abusive spouse. Still others simply opt out, vanishing in spirit or body, leaving their children alone.

Coming to grips with the culpability of a Childlike parent can be difficult because it's often easier to feel anger toward the more dominant parent. When grown children take stock of their upbringing by a Childlike parent, they may feel angry that they weren't protected but express guilt because of their anger. Childlike parents seem so weak, after all; how can you be angry with them? It is indeed tough balancing compassion for your parents' limitations with the recognition that you suffered because of their limitations.

Jack, a thirty-five-year-old salesman, can still feel the flush on his face from his father's insults and his sore bottom from his beatings. Jack's dad was an Abusing, Perfectionistic parent who seemed to delight in tormenting his son. But when the boy would seek comfort from his mother, she would invariably tell him, "Your father's feeling a lot of pressure. He's under stress and he wants the best for you. He's just trying to make you understand."

Remembers Jack, "I think I knew that my mother could never stand up to him. In a curious way, I didn't want her to. If she stood up to him, he'd win, and I would risk losing the only supportive person I had." Jack, like many children of Childlike parents, ended up protecting his protector.

In abusive families, it can be frightening to speak the truth—that the abuse is wrong. In order to avoid a confrontation, Childlike parents make excuses for their Abusing spouses and the children accept the excuses so they can maintain some kind of lifeline—even an inadequate one.

Self-Assessment

My parent(s):

- Lived "under the thumb" of a mate or others

- Rarely stood up for themselves or me

- Feared strong emotions or new situations

- Needed me to take care of them

- Engendered guilt or pity in those around them

Summary of the Eight Styles

So long as little children are allowed to suffer, there is no true love in the world.

—ISADORA DUNCAN

In reading about the eight styles of controlling parenting and reviewing your self-assessments, you may have found that one or both of your parents have characteristics of several of the eight styles. Most controlling parents are a combination of styles, usually with one, two, or three styles predominating. To review:

Smothering: Overbearing scrutiny

Depriving: Conditional love

Perfectionistic: Pressure to perform

Cultlike: Rigid rules and beliefs

Chaotic: Unpredictability

Using: Needing to be number one

Abusing: Bullying

Childlike: Inducing guilt or pity

Certain style combinations tend to cluster. Parents who are predominantly:

Smothering may also tend to be Perfectionistic and/or Using;

Depriving may also tend to be Perfectionistic and/or Abusing;

Perfectionistic may also tend to be Smothering, Cultlike, Using, and/or Depriving;

Cultlike may also tend to be Perfectionistic and/or Using;

Chaotic may also tend to be Abusing;

Using may also tend to be Depriving, Abusing, Cultlike, and/or Smothering;

Abusing may also tend to be Depriving, Chaotic, and/or Using;

Childlike may include any of the above styles. Childlike parents also tend to seek mates who are Cultlike, Using, Perfectionistic, and/or Abusing.

Splintering of Self

Unhealthy control can fracture a child's psyche, just as too much force can fracture bones, because it causes a splintering of self—chasing away some parts, jailing others, and inflating still others. If you grew up controlled, you didn't have any way to stop the hurts and losses. That these hurts and losses came at the hands of those who said they loved you only tends to deepen your grief.

In the interviews with those who grew up controlled, I listened to so much hurt. I remember Robin, whose solace from her Using, Depriving mother came from pretending that a floor mop was a twin sister. I remember sitting with Jorge as he recalled, through tears, being locked in a small room by his Abusing, Chaotic mother, with only a little barred window through which he could see other children playing outside. I remember Roberta, toilet trained at ten months and denied naps as a child by her Depriving mother, describing how each night in bed she hummed Perry Como tunes, struggling to stay awake for her father's good-night kiss when he came home from work.

Growing up controlled brings so much trauma . . .

"I had this holocaust at both ends," said Shirley, daughter of an alcoholic father and fundamentalist Cultlike mother who banned Christmas presents after realizing that "Santa" had the same letters as "Satan." "My father was always yelling, 'The economy is failing and you're gonna die. Work, work, work!' And my mother with her hellfire and damnation."

So much sadness . . .

Rosemary, the daughter of the severely Abusing, Using mother, who has battled obesity most of her life, remarked, "People say to me, 'Rosemary, why do you have such sad eyes?' and I want to say, 'Have you ever met my mother?'"

So much loneliness . . .

"If I'd gotten a few hugs and a few moments of conversation in my childhood, it might have changed a few things," said David, the son of the Depriving parents who forbade his pursuing his great love of photography. "Eventually, I built a wall around myself and shut down."

So much weariness . . .

"In my family, I felt like an alien," recalled Sharon, the daughter of the Smothering Holocaust-survivor father. "I felt like I got beaten down again and again. I eventually got the fight beaten out of me."

So much deprivation . . .

"I feel cheated out of a basic intimacy I craved but couldn't ask for," said the daughter of a Depriving mother.

So much time feeling unloved . . .

"I was convinced that my mother didn't love me. I always felt I was some sort of accident or mistake," recalled Shari, the gifted daughter whose Depriving, Abusing mother never came to her academic award ceremonies.

"I felt as if I was taking up their time. I felt I shouldn't have been born," said Samantha, whose Depriving, Abusing mother had banished her daughter's first bouquet of roses to the garage.

"My mother's attitude was, 'If you try hard, someday you will be worthy—perhaps—of my love,'" remembered Tina, whose Smothering mother had made her wear a sign reading, "Please Do Not Feed Me."

"My father seemed to love me only when I did what he wanted," mused Herb, the son of a Cultlike, Perfectionistic corporate-climber father.

So many What-ifs and If-onlys . . .

"Would I be so timid today if my father had seen a doctor and gotten a tranquilizer?" wondered the daughter of a tense, Perfectionistic father.

The daughter of a Chaotic mother mused, "I wish Mom could have loved me and set me free instead of her weird back-and-forth love."

One man raised in a Chaotic home said, "I wish someone had told me, 'You are going to grow up, make it through this, and you are going to be okay.'"

Growing up controlled means a million moments of hurt. Since children who grow up controlled are . . .

1. Not given many essentials for a healthy development

2. Deprived of resources for healthy coping

3. Confronted with forces as powerful as "brainwashing"

. . . it is to your credit that, if you grew up controlled, you came through your childhood as intact as you are.

The Price Parents Pay

While control has clear motivations and rewards for parents, they, too, pay a price:

- Smothering parents mask their aloneness by stressing conformity, yet live one step from emotional annihilation.

- Depriving parents gain power over others by loving conditionally, yet subsist on a coldness of spirit.

- Perfectionistic parents hide their flaws by coveting superiority, yet can never cease their eagle-eyed search for flaws or for anyone who may be superior to them.

- Cultlike parents avoid doubt and invalidation by proclaiming their certainty, yet cannot escape their gnawing fears that they will be proven wrong.

- Chaotic parents keep others off balance by being volatile, yet live without a safe emotional harbor.

- Using parents try to escape their emptiness by feeding off others, yet never fill their emotional void no matter how much they take from others.

- Abusing parents temporarily discharge the emotions they cannot handle by attacking others, yet sit atop a barely contained volcano of rage and guilt.

- Childlike parents escape others' demands by lowering others' expectations of them, yet feel tiny in a world of giants.

How to Accelerate Your Healing

For many children of controlling parents, a valuable step in healing is to talk about what happened to you. By telling your story, you air long-suppressed feelings and give meaning to experiences that may have left you feeling meaningless.

The stories in the preceding chapters that struck a chord in you can serve as starting points in helping you discuss your own experiences. After a childhood of not being able to speak up, it's important to find safe environments in which to talk about what happened. By doing this, you give yourself what your childhood lacked: the chance to talk without interruption, the experience of being seen and heard as you really are, and the opportunity to feel validated. Here are some possible ways in which to tell your story:

1. With a trusted friend or partner, find an hour or so when you can comfortably talk. Ask your friend to just listen. Tell them that you don't need them to solve a problem for you; rather, you'd just like to get something off your chest. Tell them to save any comments until you've finished talking. Start off by answering the question: "What makes me say I grew up controlled?"

2. Write down your experience of growing up or record it on audio- or videotape. Then read or listen to it without judgment and with compassion.

3. Find a therapist or counselor you trust and tell them you'd like to talk about your childhood.

Resources for Naming the Problem

Forward, Susan, with Craig Buck. *Toxic Parents: Overcoming Their Hurtful Legacy and Reclaiming Your Life.* New York: Bantam Books, 1989.

Golomb, Elan. *Trapped in the Mirror: Adult Children of Narcissists in Their Struggle for Self.* New York: William Morrow, 1992.

Love, Patricia. *Emotional Incest Syndrome: What to Do When a Parent's Love Rules Your Life.* New York: Bantam Books, 1991.

Miller, Alice. *The Drama of the Gifted Child*, rev. ed. New York: Basic Books, 1994.

Shengold, Leonard. *Soul Murder: The Effects of Childhood Abuse and Deprivation.* New York: Fawcett, 1989.

Next: Part Two—Understanding the Problem

The first part of this book, "Naming the Problem," described the many forms of unhealthy parental control. Part Two, "Understanding the Problem," can help you break the cycle of overcontrol by understanding exactly how it works and why it can have lasting consequences. Part Two will also explore the connections between childhood control and problems in adulthood. Making these connections can allow you to break the "trance" inherited from growing up controlled.

◆

Understanding the Problem

10

HOW OVERCONTROL WORKS

When you solve a mystery, you destroy its power over you.

Discovering how a magician does a trick can clear up your bewilderment. Analyzing advertisements to uncover methods of persuasion is an eye-opening exercise for schoolchildren. Seeing how a con artist cheated you lets you protect yourself in the future. Former members of destructive cults often get their lives back by understanding exactly how they were recruited and indoctrinated.

The same goes for overcoming the effects of growing up in a controlling family. Understanding how and why a parent controlled you can disarm both past and ongoing controls.

Part Two will help you answer the following questions:

- How does parental control operate years later despite your best efforts to get free?

- How does your controlling-family background connect to problems in your life today?

- Why did your parents overcontrol you?

Parental overcontrol is a potent, pervasive process akin to brainwashing. Controlling-family brainwashing has three components:

1. Twelve kinds of unhealthy control I call the "Dirty Dozen"

2. Distortions of responsibility through "Truth Abuse"

3. Manipulations similar to the thought-reform techniques used by destructive cults

The result of controlling-family brainwashing: Children internalize their parents' judgments and biases in the form of harsh internalized parents—those inner critics who slow our efforts to heal and grow.

The point of exploring this "brainwashing" process is not to recite the litany of our fathers' and mothers' "sins." Our parents had reasons for acting as they did. As we will see in Chapter 14, many controlling parents were emotionally wounded as children and received little compassion from their own parents. Because of that, controlling parents often have insufficient compassion to give themselves or their children.

Exploring controlling-family brainwashing is based on two paradoxes of healing:

1. To let go of a painful past, you may temporarily need to get closer to it.

2. To take greater control of your own life, you may need to revisit the days during which you had the least control.

Each of us has two sets of parents: our actual, physical parents and our internalized parents. An important part of healing from a controlling upbringing has to do with forging a healthier relationship with your actual parents. (Part Three, "Solving the Problem," will show you many options for doing that.)

But an equal if not greater part of healing comes from forging a healthier relationship with your internalized parents. I use "internalized parents" as a construct to symbolize the negative self-judgments, self-image, expectations, and viewpoints we unwittingly adopted during our years of growing up controlled. Your internalized parents are psychic stowaways, roaming through your psychological and emotional hallways and creating havoc. The better you can identify your internalized parents, the more you can take charge of your present and control your future.

Just as our actual parents brainwashed us as children, our internalized parents put us in trances as adults. They whisper *You're no good*, and we believe it. They say *You can't*, and we don't try. They urge us to

Go ahead; then, if things turn out badly, they tell us *You shouldn't have,* and we lose confidence.

You cannot change your actual parents nor can you change your past. But neither your actual parents nor your past history dictates your future. What can dictate your future is your internalized parents' judgments and expectations about yourself, others, and life in general. These judgments are nothing more than bad habits. Stubbornly rooted habits, perhaps, given the power of controlling-family brainwashing. But like any habit, they can be altered with work and time. The good news is that because they are inside you, changing your internalized parents is entirely up to you.

As we move farther into Part Two, some readers may experience that resurgence of family-loyalty feelings I mentioned earlier. Remember, by looking underneath the surface of your parents' control, you are unwrapping the myths and mysteries of your childhood. Controlling families tend to discourage such independent thinking. Even years after childhood, examining your family's overcontrol can trigger early injunctions and feelings of betrayal. For some readers, this may be the time to refer to the list of the Top 10 Guilt-Inducing Family-Loyalty Thoughts on page 10.

For other readers, exploring your family's early control may rekindle feelings of having little control over your destiny. But recognizing your internalized parents and understanding how they came to take up residence in your soul can give you the knowledge and power you didn't have as a child. Also keep in mind that your actual parents have far less hold on you today than when you were a child. There is a lot you can do to free yourself from unhealthy parental influence; much of it you have undoubtedly already done.

1. The "Dirty Dozen" Methods of Overcontrol

In the name of motherhood and fatherhood and education and good
manners, we threaten and suffocate and bind and ensnare and bribe and
trick children into wholesale emulation of our ways.

—JUNE JORDAN

Parents can overcontrol their children in twelve powerful ways, some direct and obvious, others harder to spot. I call these the "Dirty Dozen" because when they are excessively applied, they run counter to what children need for healthy development. The Dirty Dozen:

Food control

Body control

Boundary control

Social control

Decision control

Speech control

Emotion control

Thought control

Bullying

Depriving

Confusing

Manipulating

You might notice which of the Dirty Dozen listed in the following chart were prevalent in your upbringing.

The "Dirty Dozen" Methods of Unhealthy Parental Control

Method	Examples	Potential Consequences
1. Food Control	• Dictating what, when and how children eat • Dominating the dinner-table environment	• *Decreased autonomy* • *Increased emotional problems* • *Risk of eating disorders or addictive behavior* • *Poor self-image*
2. Body Control	• Excessive monitoring of body functions • Attempts to dictate dress and personal grooming	• *Violations in sense of self* • *Diminished free will* • *Risk of distorted body image*
3. Boundary Control	• Micro-managing children's sleep habits, household duties and play time • Violating children's privacy	• *Increased dependency* • *Decreased emotional safety* • *Feeling always under scrutiny* • *Lowered expectations*
4. Social Control	• Interfering in choices of friends and dates • Discouraging contact with non-family members	• *Slowed individuation* • *Distrust, gullibility or distorted ideas about relationships and other people* • *Lack of awareness of others' values*

Method	Examples	Potential Consequences
5. Decision Control	• Dominating school, career and major life choices • Second-guessing or ridiculing children's choices	• *Slowed development of "decision muscles"* • *Overreliance on parents' views* • *Increased self-doubt* • *Ambivalence over achievements*
6. Speech Control	• Dictating when and how children speak • Compulsively correcting grammar or forbidding certain words • Prohibiting dissent or questions	• *Reduced initiative* • *Slowed development of communication skills* • *Bottled-up feelings* • *Reduced confidence*
7. Emotion Control	• Overriding, dictating, ridiculing or discounting emotions	• *Reduced opportunities to learn how to cope with emotions* • *Distorted ideas about how to express emotions* • *Disconnection from a precious source of information about one's self* • *Confusion or intolerance when faced with other's strong emotions*

Method	Examples	Potential Consequences
8. Thought Control	• Attempts to regulate morals, values and tastes • Parental philosophies of life delivered as dogma • Overzealous attempts to discourage new ideas	• *Slowed intellectual growth* • *Focus on who's right and who's wrong rather than on curiosity and learning* • *Reduced self-esteem* • *Lack of awareness of other views*
9. Bullying	• Physical or sexual violence or harassment • Verbal or emotional abuse • Intimidation • Prohibiting children from defending themselves	• *Feelings of isolation and abandonment* • *Increased risk of depression and anxiety* • *Assumption that abuse is deserved* • *Poor impulse control* • *Risk of addictive behavior*
10. Depriving	• Withdrawing affection and attention when displeased • Withholding warmth and encouragement • Depriving of safety and belonging	• *Feeling unlovable* • *Increased dependency* • *Reduced confidence* • *Lowered expectations* • *Greater willingness to accept mistreatment* • *Increased risk of depression and anxiety*

Method	Examples	Potential Consequences
11. Confusing	• Unclear rules, mixed messages, erratic behavior, or baffling communication	• *Increased second-guessing of self* • *Feeling isolated* • *Hypervigilance and anxiety* • *Difficulty making decisions or taking the initiative*
12. Manipulating	• Shaming, scapegoating, and a host of other disingenuous techniques	• *Distrust of others* • *Feeling valued for appearances instead of for one's self* • *Internalization of family worries that are not children's to solve* • *Depression and rage*

Some examples of the Dirty Dozen among those I interviewed:

1. Food Control

Controlling parents' styles are often reflected in how they approach food. Using parents, for example, see dinner as *their* hour. One Using father demanded that dinner be on the table at five with the TV rolled in so he could watch the news; nobody was allowed to talk. Another Using father talked incessantly about himself, his day, and who was trying to take advantage of him at work. His captive audience wasn't allowed to leave the table until he finished.

Cultlike and Perfectionistic parents often have dinner table rituals that must be followed to the letter. One woman recalls, "Every night we girls had to get food on the table on time. My father would be shouting at us, sometimes hitting us." By adolescence, she had developed anorexia: "I felt like a pawn on a chessboard. Refusing to eat felt like the last bit of control I had over my life."

Smothering parents pressure their children to mimic parental tastes. One daughter remembers sobbing over her lawyer father's "litigation about what kind of breakfast cereal I was supposed to want."

One Abusing father literally shoved unfinished food down his children's throats. Another maintained a standing rule that if his children didn't eat "enough" vegetables, they would be forced to eat twice the normal helping. "We never knew what the right amount was, so we always had to take enough to cover ourselves," recalls his son.

2. Body Control

One Perfectionistic ex-marine father put his ten-year-old daughter on a calisthenics program of catching a football, boxing, casting, rowing, and shooting baskets—and became furious if she didn't excel.

One Abusing mother enforced a nightly ritual her son calls "the concentration camp of the dressing." He was ordered to bring all his pants to her bedroom and lay them out on her bed so she could pick the pair he would wear the next day. Then shirts. Finally, shoes and socks. If he dawdled, she would hit him with her scissors.

One Abusing, Smothering mother cleaned her daughter's ears with a bobby pin wrapped in toilet paper, invariably poking and hurting her.

Several parents seemed fixated on giving their children enemas, holding them down as the children cried.

3. Boundary Control

One early-rising Abusing father would wake his children at five A.M. on Saturdays with blaring country music. If they didn't stir, he'd crash into their rooms and throw ice water in their faces.

One Cultlike, Using father removed all the locks from his children's bedrooms and the bathrooms so he could enter the rooms at will. He declared, "It's my house and I can open any doors I want."

4. Social Control

One Depriving, Childlike mother always kept the family's drapes closed, and found it "inconceivable" that their daughter would want to leave home to see friends. The mother had few friends, yet harbored a constant fear of social ostracism. Her daughter recalls, "My mother was always saying, 'What would the neighbors think?' I could never understand why she placed so much value on the opinions of people she rarely saw."

One Smothering, Childlike mother would not allow her daughter to visit friends unless she got permission two days in advance. She also kept her teenage son from extracurricular school activities because "something might happen."

Such social isolation weakens children's autonomy. Children seek neither peers as friends, nor adult figures as mentors, for several reasons: They think nobody would want them; they want to be loyal to their parents; they are told to be self-reliant; or they know their parents simply won't allow any competition. Lacking an outside reality check, many controlled children have no way to know that they are not alone in their suffering.

5. Decision Control

One Perfectionistic, Using father conditioned his son from toddlerhood to be a doctor: "He never let it rest until I flunked out of premed. Then he told me to go to business school so I could make lots of money to compensate for the time I'd lost preparing to be a doctor."

A Using mother asked her sixteen-year-old daughter what she wanted to be. When her daughter said she wanted to be an artist, her mother said, "You can't. Only a few make a living at it. Think of something more sensible."

Decision control can be especially painful. There are few wounds deeper for children than having their dreams discounted, being told, in effect, *We don't believe in you*.

6. Speech Control

One Cultlike, Perfectionistic father habitually corrected his nine-year-old daughter's spelling and grammar in her love notes to him.

One Cultlike mother forbade her children to say "Who cares?" or to call their athletic shoes "sneakers."

One Depriving mother did not allow her son to ask for anything when they were shopping. "We could never say, 'Please buy me this,'" he remembers. "My mother thought it rude."

7. Emotion Control

One Cultlike, Perfectionistic family's rule was *Never show your feelings*. "When I was in junior high I was so unhappy, but I felt like a crybaby because I was not able to hold it in," admits their son.

One Using father devalued feelings. "He'd say, 'You can't buy any-thing with feelings. Can you touch them, see them? No. What good are they?'" his daughter recalled.

Controlling parents' mantras about emotions include:

Who cares how you feel? Just do it.

Bite the bullet and move on.

Keep crying and I'll give you something to cry about.

Don't lose control.

8. Thought Control

One Smothering mother was horrified when her fourteen-year-old son came home with Elvis Presley's "Hound Dog" record. "A young boy shouldn't be listening to that kind of thing," his mother told him. She confiscated the record and exchanged it for a Mitch Miller tune.

One twenty-five-year-old woman told her Using, Depriving mother during a car ride that she was thinking of seeing a therapist: "Once I breathed the word 'therapy,' my mother screamed and lec-tured for the whole car ride."

9. Bullying

Being hit even once can traumatize a child. One Depriving, Abus-ing stepfather flew into a rage and slapped his sixteen-year-old step-daughter. "People had always been very gentle with me, so it was very traumatic," she says. "I always lived in fear after that."

One Cultlike military father frequently berated his son in public: "It was the worst when he drank gin martinis. He'd start screaming and call me a 'stupid fuckup.' My mother would meekly try to get him to stop, but it never worked."

One woman still has a vivid mental image of herself thirty years ago sitting in the kitchen with her Abusing father standing over her, his fists clenched and ready to strike, because she was laughing and he couldn't hear the TV.

10. Depriving

One Depriving father seemed so disinterested in his daughter's wants and needs that she feels, "I could have been a cardboard cutout of myself and my dad would have treated me the same. He would just not listen. After a while, it makes you doubt yourself."

Children sense that when parents share their own stories, they share a vital part of themselves. Yet many controlling parents, closed books to their children, are unwilling to talk. Some speak only of a

romanticized past; others, only about shallow details or petty grievances. In fact, several people I interviewed knew little about their parents' pasts. In some cases this is understandable. Parents may be
ashamed of something or were raised in families that discouraged self-
disclosure. If a parent's past was traumatic, it may be hard to talk about
it. By keeping their pasts hidden or by distorting the facts, parents
unwittingly magnify the child's natural tendency to form a larger-than-
life view of them.

11. Confusing

One Chaotic mother, whose husband molested their daughter,
would first be empathic toward her daughter, then blame her for being
sexually abused, screaming, "How could you do this to me?"

One Smothering mother told her teenage daughter she should
marry late because people who marry young are not mature enough to
choose wisely. When the young woman reached twenty-five, her
mother urged her to "start seeing somebody seriously now." But, she
added, "Until you marry, you're still going to be a child." This is a classic bind: The daughter is told she'd better hurry and get married but,
until she does, she's still a child, and therefore not mature enough to
marry.

One Perfectionistic father accidentally kissed his fourteen-year-old
daughter on the lips. She recalled, "He was horrified and ran upstairs
shouting, 'Watch out for your sexuality.'" Though the daughter had
done nothing wrong, her Using stepmother called her "Hot Lips" for
months despite the daughter's tears and pleas to stop.

12. Manipulating

One Depriving, Chaotic mother blamed nearly everything on her
absent spouse. "If I tripped on the way to the bus stop, in my mother's
eyes it was my father's fault," recalls her daughter. The mother waged
war against her estranged husband by refusing to feed her daughter
dinner until she wrote letters to the father's employer saying how bad
he was. Eventually the father was fired.

One Using father's college tuition checks for his daughter came
with the written reminder, "Putting you through college is really lowering my standard of living."

Another Using father wrote on every gift he gave, "I hope you realize just how lucky you are to get this."

One Chaotic mother sent her daughter a fifty-dollar check each
Christmas, then complained about how poor she was. When the

daughter didn't cash her mother's check one Christmas to help save her money, the mother complained that the uncashed check was messing up her bookkeeping.

Self-Assessment

One or both of my parents frequently used:

- Food control
- Body control
- Boundary control
- Social control
- Decision control
- Speech control
- Emotion control
- Thought control
- Bullying
- Depriving
- Confusing
- Manipulating

Of course, reasonable control of children's behavior is necessary. But many of the Dirty Dozen methods of control, though seemingly innocuous when seen as a single instance, may very well have happened thousands of times in your childhood. Through repetition, they formed a powerful pattern that makes up the first component of controlling-family brainwashing.

The second component: "Truth Abuse."

2. "Truth Abuse"

Emotional abusers use guilt the same way a loan shark uses money: They don't want the "debt" paid off, because they live quite happily on the interest.

—Andrew Vachss

The Dirty Dozen are psychic Post-it notes by which parental views get inserted into children's minds before children develop the critical judgment to question them. Because of this, merely thinking critical thoughts about parents may cause those who grew up controlled to feel disloyal, often without knowing why.

The last two methods of the Dirty Dozen—confusion and manipulation—make up "Truth Abuse." Truth Abuse can be at the heart of the lingering feelings of disloyalty for many who grew up controlled. The effects of Truth Abuse may still cause you confusion, especially in situations concerning your rights and boundaries. Truth Abuse has several forms.

Mixed Messages

Many controlled children grow up with a stream of mixed messages—statements or actions that simply don't add up. Mixed messages can put children into inescapable binds.

For example, one Chaotic, Abusing father would rage and hit his children, then a few minutes later act as if nothing had happened and take them for ice cream.

One woman recalls her Perfectionistic mother's approach: "I was supposed to be spontaneous yet controlled."

Another describes her Smothering parents' philosophy as "Be assertive except to us."

Says yet another, "They told me not to be afraid of things, then terrorized me."

One man's Using, Perfectionistic father commanded his son never to question parental authority, then ridiculed him for not standing up for himself "like a man."

Mixed messages leave children wondering, *Can I trust what I see and feel?* As a result, they either give up or try even harder to fulfill their parents' unattainable demands. Either option weakens a child's sense of self while accentuating parental power.

Two-Faced Behavior

Controlling parents often tell their children to act one way, then act in opposite ways themselves. One Depriving, Using father presented a "saint persona" to outsiders but was uncaring at home. "He was super-polite and acted interested in strangers," his son claims. "He'd offer to give rides to my friends but he'd make me take the bus."

When parents are two-faced with others, their children naturally wonder whether their parents are being two-faced with them. The inevitable, haunting question: *If Mommy and Daddy are nice to people when they are around but trash them behind their backs, what do they say or feel about me when I'm not around?*

Dysfunctional Communication

Some children grow up in families rife with dysfunctional communication: unfinished thoughts, non sequiturs, repetitive phrases, or incomprehensible language.

Actually, much of a family therapist's work is helping families alter dysfunctional communication patterns. Dysfunctional communication focuses primarily on who is right and who is wrong, who wins and who loses, who gets hurt and who avoids pain.

In *Conjoint Family Therapy*, family therapist Virginia Satir gives examples of dysfunctional family communication (79–93) such as:

1. **Fuzzy or incomplete thoughts** like, "He's very, well, you know" (He's very what?) or, "As you can see, it's obvious" (What's obvious?).

2. **Distracting or stonewalling** when asked for clarification by saying, "You know perfectly well what I mean," "You heard me," "What's the matter, don't you understand English?" or simply restating what was said.

3. **No-choice choices** such as, "I want to go to the park, don't you?"

In most controlling families, there is an intricate system of confusing communication. One woman arrived for our interview with a flowchart she'd created that mapped out the pattern of confusing communication with which she grew up. She recited the progression of her Depriving, Chaotic parents' tactics as follows:

"If I asked questions or disagreed with them, they'd interrupt me with stock phrases like, 'You always, you never.'

"If that didn't work, they'd distort what I said. They'd become a broken record and turn the focus back on me, saying something like, 'Interruption surely is a big obsession with you.'

"If I persisted, they'd try to distract me by bringing up some totally separate side issue. Or they'd make guilt-provoking statements, throw a temper tantrum, or label me a 'paranoid schizophrenic.' If I walked away, it would only intensify the process at a later date.

"The net result was that I'd end up responding to their attacks and lose track of what I really wanted to get across."

Unclear communication blocks children from expressing all the emotions and ideas they have bottled up inside, causes them to doubt their perceptions and communication skills, and makes them feel invisible.

Outright Denial

Controlling parents can create uncertainty simply by denying things a child knows to be true. Pat Conroy, author of *The Great Santini* and *Prince of Tides*, wrote in the introduction to Mary Wertsch's *Military Brats* that parents can turn children into "unwitnesses of our own history" by changing stories, challenging their children's experience, or denying their children's memories (xix).

For example, when Elizabeth, a thirty-one-year-old travel agent, began therapy in her mid-twenties, she realized how depressed she had been as a child. When she told her Perfectionistic, Using mother of her realization, her mother said, "What are you talking about? You had a great childhood. You were so happy and joyful."

Elizabeth: "A lot of the time I think I'm crazy when my mom disputes my version of things. Her beliefs are so strong they kind of just wipe out mine. She'll say I had a happy childhood and I'll think she's right." Perhaps Elizabeth's mom never noticed her daughter's depression. Or perhaps she would have felt guilty if she were to admit that her daughter had been depressed.

Julia, now a twenty-six-year-old clerical worker, remembers being confused at fourteen by the icy dinner table atmosphere between her Depriving mother and new stepfather. One night she dared to ask, "Isn't it sort of ridiculous that we just sit here and never talk?" Her words were met with stony silence and looks of disgust indicating they thought she was crazy. Julia remembers thinking, "Well, I thought what I said was true but they think I'm crazy. So either they're lying or I'm wrong. They wouldn't both lie, so I must be wrong."

Julia's dinner table experience is at the heart of how confusion becomes control. When faced with conflicting messages, children can conclude that either:

   a. My parents are wrong, lying, or out of control.
   or
   b. I must be wrong or missing something.

Most children invariably conclude that they are wrong or missing something. They don't want to think that their parents are out of control—even if they are—because it would leave them feeling as if they're in a car hurtling down a mountain with no brakes and no one at the wheel. By doubting their own perceptions and giving the bene-fit of the doubt to their parents, children try to reduce their terror. Instead of feeling hopeless, they feel hopeful that perhaps their parents are right, and that nothing really is wrong. Instead of feeling helpless, children can pretend they have at least some power by thinking, *If I just listen more closely, watch more carefully, and try to be good, my par-ents will treat me better.*

Unfortunately, by adopting misperceptions to reduce their hope-less and helpless feelings, children only increase their guilt and self-doubt. Remember Rosemary, whose Abusing, Using mother once yanked a red bow out of her hair and called her a "whore" for wearing it? "My mother had me completely brainwashed," she maintains. "She had me lying about the bruises she inflicted on me, saying I walked into a door, and I actually believed it."

Controlling parents are generally disinterested in exploring their adult children's grievances with them. When one forty-four-year-old man wrote a letter to his self-centered father telling him what trou-bled him in their relationship, the Using father wrote back with an attack. "He said, in essence, 'Don't blame me—everything good in your life you got from me, everything bad in your life is your own doing,'" recalls his son.

Emotional Dumping

Many controlling parents spill emotional angst in their children's laps, then walk away. The parents may feel relief but their children feel anything but.

Jack, a thirty-five-year-old salesman, remembers how his Childlike mother would hysterically fret and fume to him about real and per-ceived slights from others. When her son suggested actions she might take, she'd "yes, but" him. Finally she'd undo it all by saying, "Well, it's not worth getting angry about."

According to Jack, "I'd walk away feeling upset and never knew why. She'd plant all this anger, then say, in effect, 'Never mind.' Maybe she felt better afterward, but I'd end up all stirred up with no place to go." Jack couldn't win. When he suggested solutions, his mother found reasons for why each action wouldn't work. When he invested time in listening to her, she discounted her anger and, by implication, his efforts to help, saying it wasn't worth getting angry about.

When parents dump, their children tend to focus on parental needs instead of their own. Dumping can engender a lasting fear that anyone you love will emotionally engulf you.

Intimidation

Many people I interviewed vividly described "the Look" or "the Voice," those stern glances or threatening tones they got when doing something "wrong."

"My mother exercised an amazing amount of control just with looks," Samantha, the forty-year-old artist, told me. She remembers giggling with her sister while watching a play, then freezing when her Depriving, Abusing mother give her "the Look": narrowed eyes, pursed lips, and an almost imperceptible shake of her head: "It was a silent kind of control. I was completely under her spell."

Roberta, the fifty-seven-year-old homemaker, demonstrated her Depriving mother's "Look" just so: a slight tilt of the head, a sternly raised eyebrow, and an uplifted pointing finger: "She didn't even have to raise her voice. I was so tuned in to her. I'd freeze on the spot."

Others described the threatening tone of voice their parents took when they wanted to control their children. "A growl bristling with contempt," one man said of his Using, Abusing father's control voice. Many say they still cringe involuntarily when their parents assume "the Voice," even on the phone. "I can recognize it from the very first word he says," one woman admits. "I never know what causes it, but I become like a little girl and just want to curl up."

"The Look" and "the Voice" can cause children to feel as if they have lost their parents' love.

Other Techniques of Truth Abuse

Scapegoating: Labeling one child as the source of all family problems. Scapegoating is a distraction that hides parental responsibility. It also conveys the message that children have no control over where blame will fall; it is totally up to the parents.

Attaching emotional strings: Equating money with love or giving gifts with the expectation of getting something in return. Attaching emotional strings may lead children to overvalue material goods or, conversely, to develop a love-hate relationship with money since it was used as a substitute for affection.

Infantilizing: Reducing children's stature and self-confidence by treating them like infants when they no longer are. Infantilizing parents talk to their grown offspring as if they are still children, pick out clothes for them, or order for them at restaurants.

Parentifying: Forcing children to take on adult responsibilities before they're ready. Says one woman, "I was never a little girl. I was my mother's mother." Robbed of adequate time to be young, parentified children are set up to take care of others at the expense of their own needs.

Erratic behavior: Mercurial moods and unpredictable, dramatic behavior that gives parents the freedom to act however they want. Faced with this, children wonder what they did to spark it. Hoping, *If I can figure out what I did to cause their reaction, I can stop it*, children grow up second-guessing and blaming themselves.

Projecting: Unconsciously attributing to another person the distressing qualities or feelings that you don't want to experience yourself. By so doing, parents lead children to look inward for blame and to become confused about what is real.

Triangulating: Unfairly involving children in marital matters, such as by confiding marital problems behind a spouse's back. This gives children a distorted sense of their stature in the family and can leave them saddled with guilt.

Martyrdom: Playing the role of martyr by using phrases such as: "If it weren't for you," "I do so much for you," "This is your fault," or "How dare you!" Martyrs tug on children's innate love for their parents. Sometimes the parents really are in need, but children can never be sure whether the need is real or not. Rather than let their parents down, they feel compelled to do a dance of caring anytime a parent plays the martyr. When parents violate boundaries in this way, children's abilities to differentiate between their own problems and those of others may be handicapped.

The Results of Truth Abuse

When children are baffled by mixed messages, they often conclude that they're stupid or must have missed something. When parents live a double standard, children conclude that they are not as important as their parents, perhaps not as important as anybody else. When parents deny children's pain, children see their own perceptions as faulty. Above all, they learn not to trust themselves.

Confusing and manipulating children is treacherous to their development. When parents scapegoat, they twist children's experiences of the truth. When they dump, they twist children's feelings. When they infantilize, parentify, or triangulate, they twist children's relationships with others. When they become martyrs, they twist children's relationships with the parents. When they project or shame, they twist children's relationships with their very selves.

Confusing and manipulating allow parents to strengthen their one-up position. In many cases, controlling parents lack social skills, communication skills, and a tolerance for ambiguity. Instead of adapting to change, which is crucial for parents since children are constantly changing, controlling parents simply do or say what they have always done or said, whether it works or not. By their actions, parents put their children under a microscope and avoid facing scrutiny themselves.

In adult life, the legacy of Truth Abuse is felt especially in our personal relationships. For example, we may seek roundabout ways of making our feelings and needs known, then be upset when others miss our cues. We may expect others to have hidden agendas. We may be slow to assert or protect our rights. We may feel confused when we are wronged, then look to the wrongdoer for guidance about how we should respond. Each of these responses is a relic of controlling-family brainwashing. They are, to our detriment, our internalized parents playing fast and loose with the truth.

Self-Assessment
One or both of my parents frequently used:

- Mixed messages

- Two-faced behavior

- Dysfunctional communication

- Outright denial

- Emotional dumping

- Intimidation

- Scapegoating

- Emotional strings

- Infantilization

- Parentification

- Erratic behavior
- Projection
- Triangulation
- Martyrdom

Truth Abuse and the Dirty Dozen are the first two components of controlling-family brainwashing. The final component: manipulations similar to the thought reform used by destructive cults.

3. Cultlike Manipulations

My mother had me completely brainwashed.

—ROSEMARY, 55, A MANAGER

While households are not cults, and you were not a cult member (though it may have felt like it), many controlling families share essential characteristics with cult indoctrination and thought reform. In destructive cults, members forfeit their bodies, minds, and lives to the cause. In controlling families, children forfeit their autonomy, development, and spirit.

One woman freely admits that her father's control made her feel as if she lived in a cult: "There was one dictator and all decisions had to go through him. We were isolated from other beliefs and values and there was a sense of betrayal if you talked to others. If you disagreed, you'd get ostracized. It was, 'Obey or we will not talk to you.'"

In a process that is remarkably similar from cult to cult, recruitment and indoctrination of cult members works by manipulating five areas: feelings, behavior, thinking, relationships, and identity. Strikingly, controlling families exhibit parallels in each of these five areas.

Parallels Between Destructive Cults and Controlling Families

Destructive Cults	Controlling Families
1. **Manipulations of Feelings** • Cults give members approval alternating with appeals to fear and guilt • Leaders ridicule members' emotions that conflict with cult goals while rewarding members' emotions that support cult goals	1. **Manipulations of Feelings** • Parents give approval when pleased but withhold affection when displeased • Parents ridicule or forbid children's "unacceptable" emotions such as anger, sadness, or fear while exhorting children to be "proper"
2. **Manipulations of Behavior** • Behavior is rigidly proscribed through control of sleep, diet, privacy, dress, access to information, activities, and relationships • Sensory overload (chanting, singing, meditating, lectures, speaking in tongues and/or testimonials) dulls members' senses • Cults encourage members to inform on one another • Cults stress compliance to cult rules and rituals that, no matter how mundane or odd, must be followed to the letter • Questions are shamed or avoided and the focus turned on the questioner	2. **Manipulations of Behavior** • Parents control children's sleep, diet, privacy, dress, access to information, activities, and relationships • Excessive chores, lectures, repetitive clichés or family rituals keep children preoccupied • Parents scapegoat and play children off against each other • Parents stress compliance to rules and rituals that, no matter how mundane or odd, must be followed to the letter • Parents silence disagreements by labeling dissent as a "sin"

Destructive Cults	Controlling Families
3. Manipulations of Thinking • Cults profess freedom and openness but foster dependence, restricted information, and lack of intellectual rigor • Cult credo and needs supersede individual needs or desires • Black-and-white, all-or-nothing thinking pervades • Cult credo generally cannot be explained or disproved and is said to be fully understood only by a select few	**3. Manipulations of Thinking** • Parents foster "Truth Abuse" by denying their destructive actions and being unwilling to discuss them even years later • Parental needs, morals or relationships are seen as all-important • Parents have little tolerance for the gray areas in life • Parents confuse their children with mixed messages or simply answer, "Because I say so"
4. Manipulations of Relationships • Cults provide "instant intimacy" at the price of insisting that members reveal innermost thoughts, feelings, and habits • Relationships with outsiders are discouraged by fostering an "Us vs. Them" mentality • Members forfeit financial, social and emotional resources and give cult leaders the right to make personal decisions for them • Leaders are seen as possessing unique goodness and lacking faults or insecurities	**4. Manipulations of Relationships** • Parents violate children's privacy by searching rooms, opening doors without warning, or eavesdropping • Families tend to be socially isolated, jealously guard "family secrets," and harshly judge "different" types of people • Parents feel they own their children and can treat them as they like • Parents rarely admit their mistakes

Destructive Cults	Controlling Families
• Leaders are accorded special rights, privileges and living conditions • Leaders demand that members follow grandiose schemes as a test of loyalty	• Parents treat their children as second-class citizens • Parents see children's desires for independence as a rejection of parents
5. **Manipulations of Identity and Sense of Self** • Happiness is seen as flowing from the group and leader while unhappiness is seen as flowing from within members • Actions that distance members from the cult bring pain while actions that move members closer to the cult bring pleasure • Leaders act as sole judge of worth, truth and behavior, with the right to punish and reward	5. **Manipulations of Identity and Sense of Self** • Families are organized to protect and serve the parents, not to optimize individual growth • Children feel disloyal when acting or feeling different than parents • Parents criticize their children's character or nature, rather than their actions

One or both of my parents frequently tried to manipulate my:

• Feelings

• Behavior

• Thinking

• Relationships

• Identity and sense of self

The Dirty Dozen, Truth Abuse, and cultlike manipulations make fertile ground for the harsh inner critics who can control you for years after you leave home. The next chapter will help you uncover more about your inner critics' genealogy.

11

PUTTING IT ALL TOGETHER
Meet Your Internalized Parents

Things we can see through do not make us sick, although they may arouse our indignation, anger, sadness or feelings of impotence. What makes us sick are those things we cannot see through.

—ALICE MILLER

The elements on the left-hand side of the chart below are ingredients for optimal development. The elements on the right are a prescription for slowed development and unfulfilled potential. You might notice where your family fell on each of these continuums:

Prescriptions for Optimal and Slowed Development	
Rx For Optimal Development	**Rx for Slowed Development**
Safety	Stress
Autonomy	Dependency
Love	Deprivation
Respect	Attack
Attention	Neglect
Connections with Others	Social Isolation
Learning Experiences	Neglect of Learning
Avenues for Self-Expression	Blocked Self-Expression
Accurate Self-Image	Distorted Self-Image
Healthy Interpersonal Boundaries	Unhealthy Interpersonal Boundaries

In human development, the whole is greater than the sum of its parts. For example, if your upbringing swung to the left-hand side of this continuum—the prescription for optimal development—the effects were cumulative:

If you felt safe and nurtured, you were more likely to express yourself and connect with others . . .

. . . the more you could express yourself and connect with others, the more you could respect yourself . . .

. . . the greater your self-respect, the more likely you were to act with autonomy and initiative . . .

. . . acting with more autonomy helped you to foster healthy boundaries and an accurate self-image. And so on.

By the same token, if your upbringing swung to the right-hand side of the continuum—the prescription for slowed development—the effects were also cumulative:

If you grew up with attack, neglect, or deprivation, your self-image may have become distorted . . .

. . . if your self-image was distorted, your willingness to express yourself may have suffered . . .

. . . with reduced self-expression, your social isolation may have increased, leaving you more dependent on your parents . . .

. . . with enhanced dependency, you may have been more vulnerable to relationships with unhealthy boundaries . . .

. . . this greater vulnerability probably created more stress and dependency. And so on.

Controlling families exact a cumulative toll because the key avenues to mental health—access to information, supportive others, emotional expression, and free speech—are generally missing. Children in controlling families tend to lack a sympathetic adult who believes in them. Controlling families also tend to disable children's healthy natural instincts and magnify the already unequal relationship between parent and child.

Despite the sense of mystery about their upbringing that many people who grew up controlled possess, in retrospect it's not so mysterious:

The people you depended on for your survival . . .

The people who had the ability to give you tremendous pain or pleasure . . .

Controlled you in a dozen tangible ways . . .

Thousands of times . . .

In your most impressionable years.

In essence, controlling parents brainwash with a one-two-three-four punch:

1. Creating an environment hostile to growth

2. Blaming their children for creating the environment

3. Criticizing their children when the children suffer the consequences of the environment the parents created

4. Denying doing any of this

It wasn't fair.

It wasn't right.

And you do have the right to feel anger, sadness, dismay, and much more over what was done to you.

If you have doubts about the tremendous power and impact of controlling families, it can help to review their arsenal. Seeing it all together can be shocking—and freeing.

Controlling Parents' Arsenal . . . and Cost to Children	
Controlling Parents' Arsenal	**Cost to Children**
Conditional love	One-down position
Disrespect	Feel undeserving
Labeling dissent as a sin	Eroded autonomy
Long-standing family tension	Sapped energy
Lack of praise	Negative self-image
Harsh discipline	Focus on obedience, not learning
Confusing communication	Rampant self-doubts
Pervasive mistrust	Isolation
Unhealthy boundaries	Distorted sense of self
Excessive scrutiny	Increased second-guessing
Social isolation	No sources of support
Smothering uniformity	Hindered initiative
Deprivation	Lowered expectations
Perfectionistic pressure	Reduced self-acceptance
Cultlike thinking	Curtailed curiosity
Chaotic atmosphere	Diminished trust
Using parenting	Impaired coping skills

Controlling Parents' Arsenal	Cost to Children
Abuse and intimidation	Crippled self-protective instincts
Childlike parenting	Parental needs dominate
Food control	Increased dependence
Body control	Reduced pride
Boundary control	Insecurity
Social control	Heightened depression and anxiety
Decision control	Lessened free will
Speech control	Blocked self-expression
Emotion control	Narrowed resources for coping with stress
Thought control	Complicated inner life
Truth Abuse	Parental denial prevails
Mixed messages	Confusion and paralysis
Two-faced behavior	Uncertainty and mistrust
Scapegoating	Obscured parental responsibility
Infantilizing	Prolonged dependency
Parentifying	Children become caretakers
Triangulating	Split loyalties and increased guilt
Emotional dumping	Feelings of failure
Assumptions of "owning" children	Children accept abuse and control
Attacks on children's very nature	Shattered self-esteem
Distorted models of relating	Warped expectations for relationships
Black-and-white thinking	Warped intellectual development

The net result of growing up under the guns of this arsenal: **To survive, children internalize the controlling voices of parents.**

As I've said, it explains why parental control may affect you even today.

Why Your Internalized Parents Are So Powerful

There are three "givens" about parents and children:

1. Parents and children have an inherently unequal relationship.

2. Young children tend to idealize and mimic their parents, making it difficult to achieve a balanced view of them until much later.

3. Children need love, attention, and approval from their parents and will do anything to get it.

In healthier families, parents take advantage of these three givens to socialize, teach, nurture, and love so that children will grow up emotionally stronger. In controlling families, however, parents take advantage of these givens to get more control.

Young children, not yet complex thinkers, aren't able to see the grays and the nuances in life. To them, Daddy and Mommy are big and good, whether they are or not. Children can be scrutinized at any time by parents: when eating, playing, sleeping, and on the potty, but few children see their parents sleeping, making love, or using the bathroom. Few see their parents at work or in the outside world, when the parents may not be as dominant or as in control. As a result, children grow up seeing their parents as larger than life.

In controlling families, the negative influences of parents are magnified. If a parent is chronically anxious, the child—self-centered, as children naturally are—may conclude that there is something dangerous or wrong about themselves. If a parent cannot relax or gets tense even on happy occasions, a child may conclude that joy and happiness are not okay. If a parent is uncomfortable around anger, children may conclude that anger is to be feared or that their own anger will damage others. These conclusions go deep and can last a lifetime.

More than anything, children want love. When you are a helpless, tiny creature in a world of giants in which events happen that you don't understand and can't control—a "blooming, buzzing confusion," as William James called an infant's experience—a parent who loves you and whom you can trust and love is the top priority for survival. Children need not only love but also all that goes with it: nurturing touch; acceptance; safety; belonging; being seen for who they are; and the freedom to laugh, cry, rage, and be afraid. Because they need love and acceptance so desperately, children will take them in any form they can get them. When they don't get love, they'll construe what-

ever they do get—including unhealthy control—as love. Therein lie
the seeds of problems later in life.

The most unfortunate parallel between controlling families and
destructive cults is that parental control becomes internalized in children, just as cult dogma becomes internalized in cult members. No
parent can be present twenty-four hours a day. But controlling parents
don't have to physically be there because the family system installs an
omnipresent inner controller in the child. These twenty-four-hour
internalized parents, with their nagging commentary, second-guessing,
and criticism, can perpetuate deprivation, perfectionism, and speech-
and-feeling control well into adult life. This inner control may surface
in the form of poor interpersonal boundaries, feelings of unworthiness,
lowered expectations, self-loathing, fear of closeness, or poor self-
image.

Looking back, it may be alarming to see how controlled, even
"brainwashed" you were as a child. Yet, like members of cults or pris-
oners of war, you had little choice. You didn't do anything wrong. Anybody in such a closed system would have suffered. Knowing this, you
can assure yourself that:

You are not crazy.
You didn't make it up.
Overcontrol really happened.
It was painful and destructive.
You could not help but internalize controlling parental voices.

These realizations open the door for a further realization that can
pave the way for you to let go of much of the destructive legacy of
childhood overcontrol:

If you could not help but internalize controlling voices, then many of
your self-criticisms, fears, and doubts are not yours, nor are they your true
voice. They are merely messages from your internalized parents. They are
relics from a controlled past. They are simply bad habits. And you can
change them.

Exercise for Understanding Overcontrol and the Internalized Parents

Recall an encounter with a parent or any controlling person and check
off in the first column which of the Dirty Dozen control methods they
used.

Then write down any self-critical thoughts you recall having during or after the encounter. These are messages from your internalized parents. In the second column, check off the kinds of control these messages from your internalized parents *represent*.

	Parents used	Internalized parents use
Food control	○	○
Body control	○	○
Boundary control	○	○
Social control	○	○
Decision control	○	○
Speech control	○	○
Emotion control	○	○
Thought control	○	○
Bullying	○	○
Depriving	○	○
Confusing	○	○
Manipulating	○	○

12

THE ADULT-LIFE LEGACIES
OF GROWING UP
CONTROLLED

A lot of people go through life beating themselves up the same way they were beaten up.

—MARLO THOMAS

Controlling families harm for one predominant reason: They are organized to please and protect the parents, not to foster optimal growth or self-expression among family members. Such a skewed structure can distort a child's sense of self in ways that last well into adulthood. This chapter focuses on learning to make connections between early control and your present problems. Remember: Growing up controlled is not a life sentence. What you can see, you can heal. You were not to blame for being controlled nor are you to blame for the consequences of that early control. You may have been *victimized*, but you are *not a victim*. Grasping how early control relates to current life challenges can be the key to mastering many of those challenges.

You aren't responsible for what your parents did to you, they are.

You are responsible for what you do with your life now, your parents aren't.

Here are stories from three people interviewed. Each grew up in a very different type of family, but all share a commonality: Early control is linked to many of their adult dilemmas.

Alex

Alex, forty-eight, a bearish man with black curly hair and blue eyes, has been the top seller in his company for seven years running. He owns a stylish home, impeccably furnished and maintained. Despite his financial success, he finds little pleasure in life.

Alex was married briefly in his twenties, but divorced. He lives alone and is not seeing anyone. Alex's romantic relationships have taken one of two paths: He either becomes too demanding, and his partners leave, or his partners want more closeness than Alex can bear, and he leaves: "I want someone to get close to but when someone gets too close, I run." It's difficult for Alex to laugh or relax and he finds physical contact uncomfortable, stiffening involuntarily when he is hugged.

On the job, Alex is compulsive. "I hate getting phone calls on Friday afternoons at five because there is nothing I can do to take care of it until seven A.M. Monday. I want instant response, instant solutions," he admits. Alex takes as little comfort from his career successes as he does from his financial ones: "I refuse to quit my job but I almost hope I get fired so I can get rid of everything I own and go somewhere and write novels."

Alex's father died of complications from diabetes. Alex has a weight problem and he suspects that he, too, may have the disease but refuses to get checked: "I do have low energy and low blood sugar shakes at times, but if a doctor told me I had to give myself shots, I'd never do it."

Alex's Upbringing

As I listened to Alex, I wondered what led to his malaise. It turns out that he grew up in a woefully aloof family with two Depriving parents. Alex does not remember being held, hugged, or told he was loved by his parents. His parents rarely asked his opinions. "One time at dinner a relative asked me my opinion about some world event," Alex recalls. "I didn't know what to say. Our family never had opinions. We never talked at the dinner table. I went through childhood like a blob. I had learned for years not to rock the boat." Once, on an overnight high school field trip, Alex was silent the entire time until, he remembers, "One kid asked me, 'Don't you ever talk?'"

Alex's father rarely spoke to his son: "Some dads tell you that you have to hit a home run each time. My father never even said, 'Let's play catch.'"

He never recalls seeing his mother laugh or cry, felt she kept him at arm's length, and was mostly concerned that Alex "do the right thing." His mother died five years prior to our interview. Shortly before her death, Alex visited her and took a walk with her. It was the first and only walk he ever took with her in his life.

Connections Between Alex's Past and Present

The more I heard about Alex's childhood, the more I could make connections with his current struggles.

It's understandable that Alex might find touching difficult when he had little physical affection as a child. It's understandable that he might distrust closeness when he failed to gain his parents' affection no matter what he did. It's understandable that Alex sometimes pushes too hard in relationships since love was so hard to get in his family. After a childhood of never being asked his opinions, it makes sense that Alex feels unattractive to others. It's also understandable that he feels driven in his work life and finds it hard to relax. His family was a place where doing the "right thing" was more important than having healthy feelings.

And it's equally understandable, though tragic, that Alex secretly hopes to be fired and ignores a potentially threatening medical condition. Alex was never taught to love himself. Since he was rarely held or touched, he grew to feel unlovable. Since his interests were devalued, he learned not to value himself. For so many years, try as he might, Alex could not get what he wanted: his parents' approval and demonstrations of love. As an adult, Alex displays the symbols of success but lacks the substance.

Alex's challenge, despite his lack of early healthy models for loving and relating, is to cherish himself and gradually open his heart to others.

Penny

Penny has returned to school at age fifty-three to finish her B.A. in fine arts. She has two grown sons and is in her third marriage. "This time, I finally got it right," she says about her current marriage.

Penny told me she has lived most of her life behind a "Good Girl, Miss Perfect false front" and finds it hard to assert herself, giving everyone but herself the benefit of the doubt. She has trouble accepting compliments: "I'm amazed when people say I am a good listener."

Prior to her present marriage, Penny tended to pick men with whom she gave endlessly but who never met her needs. "I'd end up with people who treated me like a thing," she confesses.

Penny tends to be overly concerned with "shoulds" and "rules." She is often frightened by strong emotions and is easily startled: "Sometimes I jump when my husband, who is super gentle, comes up behind me to hug me."

Penny had fixed me tea before our interview and was bustling around picking up tea bags and wiping the countertops. She caught herself and laughed, then grimaced: "I am so compulsive. That is one thing I got from my parents. I find myself following guests around and cleaning up after them."

Penny's Upbringing

Penny recalls having little room to be herself around her Smothering mother and Perfectionistic, Cultlike father. When she was twelve, a friend of her parents asked her how she liked school; Penny told the truth: She hated it. Her embarrassed mother quickly contradicted her, "Oh, no, she *loves* school." Her mother got her to do unpleasant chores by waiting to ask her in front of others, when Penny was too embarrassed to say no.

Even in Penny's teens, her mother insisted on ordering for her in restaurants. When Penny moved to a vegetarian co-op, her mother brought her home-cooked meat dishes and refused to leave until Penny ate them.

In her first semester of college, Penny's mother asked about her major. "I don't know yet," Penny responded. "You've got to know," her mother insisted. Within a few weeks, Penny decided to major in journalism, which her mother then demeaned: "She told me I had to do something 'practical' instead."

Penny's father, an engineer and ex-marine, insisted on military-style discipline. She and her sister had to sit perfectly still at dinner until their father picked up his fork, which signaled that they could begin eating.

Penny remembers her dad constantly telling her to "Calm down." The impact of his cautions still haunts her. "Even today," she says, smiling ruefully, "if I get happy or sexually excited, I hear his 'Calm down' in my head and I often stop myself."

Her father could never understand Penny's nonlinear way of reasoning. When she explained that she behaved according to feelings and intuition, he dismissed her reasoning as wrong: "He'd call me an egomaniac who needed therapy."

"As a kid, my predominant feeling was disappointment in myself," Penny freely admits. "I always felt that if I tried a little harder, maybe someone would notice and approve. I blamed myself for not being perfect and lovable."

Connections Between Penny's Past and Present

It's understandable that Penny has lived behind a "Miss Perfect" false front. As a child, her real self wasn't welcomed. It's understandable that Penny finds it hard to trust her intuition. How can Penny feel spontaneous when her father's "Calm down" still echoes forty years later? How can Penny allow herself even justified anger when her parents seemed so rigid?

Since her parents acted larger than life, it makes sense that Penny questions herself, offers others the benefit of the doubt, and has endlessly given to romantic partners at her own expense. Penny's upbringing left her feeling like a second-class citizen, apologizing for her existence.

After growing up in a "boot camp" atmosphere, it's easy to see why Penny has an exaggerated startle response, even to her husband. Why shouldn't Penny have a compulsive streak and trouble cherishing herself when her models were a mother who left little breathing room and a father she could never please?

Penny's challenges are to give herself enhanced permission to respect herself and to have faith that life can bring positive, not just negative, surprises.

Belinda

Belinda, thirty, is a computer analyst. At work, she feels intimidated by her boss and "extrasensitive" to office politics. Though she has a good job, with seniority, she is "completely mystified" about what she wants to do with her life.

Belinda feels that she's on an emotional roller coaster. She often finds it difficult to define her desires and stick by her decisions. She feels "flawed" and has struggled with eating disorders for much of her life. When she gets stressed, she freezes. "I space out, complete with physical tension and armoring," she admits. "I turn my hostility on myself, get sloppy, overeat, and then hate myself." When Belinda gets angry, she often soon feels depressed: "I'll scream a lot, slam things, and then I'll feel such despair."

Belinda's boyfriend of two years has begun to talk about marriage, but she feels a long way from making such a commitment. In previous relationships, she has picked men who tended to dominate her. Although her current boyfriend is less controlling, Belinda sometimes feels smothered. At other times she worries that he might leave her. "I cling, I get afraid, and then there is no true intimacy. I'm confused over whether I am even able to be intimate." As with many of those I interviewed, Belinda has strong doubts about having children.

Belinda's Upbringing

Belinda was raised by a Chaotic, Using single mother. Her mother's moods changed wildly and her efforts to discipline Belinda were erratic. Curfews and rules were ignored one day, enforced the next.

Her mother often seemed downright mean. In second grade, Belinda's teacher wrote on her report card, "Belinda is very smart and her reading is way ahead of average. She's a good student, though she does have a tendency to visit with her neighbors." Ignoring the teacher's praise, her mother wrote back, "Do you want to punish her or should I? The best way to punish her is to humiliate her in front of others."

When fourteen-year-old Belinda's tennis coach said Belinda showed great promise, her mother stopped the lessons, claiming they were a luxury the family could not afford. When Belinda was fifteen, her parakeet was killed by a neighbor's cat. It was all Belinda's fault, according to her mother, who forbade Belinda from crying about it.

Her mother repeatedly told Belinda she was "ugly," "fat," "disgusting," "smelly," and a "mistake." Years later, when Belinda told her mother how much those words had hurt, the older woman brushed her off, claiming, "Oh, you must have known I didn't really mean it." Equally confusing were her mother's moods. Her mother was frequently depressed and would lie in bed for hours, shades pulled, but when Belinda asked what was wrong, her mother told her, "Nothing's wrong. I'm fine." Other times, if she asked her mother what she was angry about, she would reply, "Oh, I'm not angry, I'm just sad." Yet, says Belinda, "People who meet her say she's the angriest person they've ever seen."

While her mother's changeable behavior left Belinda's head spinning, her Abusing stepfather, who entered the family when Belinda was fourteen, related to his stepdaughter with derision. He regularly told Belinda she was "a tub of lard" and jeered, "You'd even gain weight from drinking a Tab," yet he made her clean her plate.

Connections Between Belinda's Past and Present

It's understandable that Belinda has difficulty with authority figures and office politics when the authorities and politics in her family were so erratic. It's not surprising that she is mystified about what to do with her life when she grew up baffled by her mother's confusion. It makes sense that Belinda finds it hard to desire things when anything she excelled at was taken away. Because of her mother's unclear rules, Belinda was constantly on the lookout for clues and cues. Rather than thinking that the rules didn't make sense—which would have put her in violation of controlling parents' number one rule of no dissent—Belinda concluded that she was "flawed."

No wonder Belinda has trouble knowing her own heart. Her mother's denial of her own obvious emotional difficulties, such as saying she was "fine" when she was depressed, left her daughter doubting her own perceptions. Given the tension in Belinda's early life and the coercive messages around food, it is no wonder she has faced eating disorders.

In her relationships with men, it's not surprising that Belinda has made some unhealthy choices. How could she feel good about being a woman or find a man who would treat her well when her stepfather degraded women? How could she maintain a clear sense of self when her mother flip-flopped on discipline? How could she help but struggle with intimacy when life with her mother was so unpredictable and intimacy was so elusive?

Belinda couldn't ask her mother for help since her mother was so erratic. She couldn't ask anything of her stepfather because his boundaries were so untrustworthy. She was too ashamed to ask for help from anyone outside the family. And she couldn't find answers within herself, since she had come to doubt her perceptions. Ultimately, she blamed herself.

Belinda's challenge is to build more solid emotional ground by trusting her instincts, reducing self-blame, and learning to love herself.

Potential Consequences in Your Adult Life

Like Alex, Penny, and Belinda, you may now be making connections between the unhealthy aspects of your upbringing and your present problems. As I've cautioned, seeing these links may spark sadness, anger, guilt, disbelief, or feelings of being overwhelmed. Yet confronting the connections may very well bring relief—as mysteries fall into place, and you reclaim a part of your rightful heritage.

Being raised by a controlling parent is a series of trials that meld into one years-long, slow-moving trauma that, if not attended to, can manifest delayed consequences. Among the signs of posttraumatic stress are (the items in italics are from the *American Psychiatric Association's DSM-IV* 428):

- *Acute sensitivity to internal or external cues that symbolize or resemble aspects of the trauma.* For some who grew up controlled, this can mean acute fear of others' anger, or intense negative reactions when feeling dominated or ignored.

- *Efforts to avoid thoughts, feelings, activities, people or places that may trigger recollections of the trauma.* For some who grew up controlled, this can mean avoiding situations in which they would have power over others, feel dependent, or think about their parents or past.

- *Feeling numb, detached, estranged from others, or feeling disinterested in participating in significant life activities.* For some who grew up controlled, this can mean difficulty in attaching to others in intimate relationships; apathy to love, friendships, work, or play; or feeling as if you must go it alone in life.

- *Difficulty feeling emotions outside a limited range.* For some who grew up controlled, this difficulty may manifest itself as an inability to feel anger, joy, pleasure, trust, or love. Posttraumatic stress means being constantly in fight-or-flight mode. When someone is in fight-or-flight mode, there is little room for joy, relaxation, trust, or optimism. Life becomes solely about survival.

- *Irritability, hypervigilance, or difficulty sleeping or concentrating.* For some who grew up controlled, this can mean a tendency to be easily startled, a feeling of being judged even when nobody is around, or excessive concern about personal safety or being intruded upon.

- *Reexperiencing the trauma, sometimes in the form of intrusive thoughts, images, dreams, or flashbacks of traumatic events.* For some who grew up controlled, this may mean sudden mental images of a parent in a threatening pose or controlling manner, intrusive and critical mental dialogues in a parent's voice, or the sensation of being trapped in situations similar to childhood ones.

To heal from trauma, it's essential to gradually explore the feelings, memories, and beliefs that arose from the wound. This process opens the door to releasing the emotional charge that was buried during the trauma. It's important to find safe settings in which to talk about the trauma and its effects; by so doing, you can reestablish feelings of safety and mastery. In Part Three we will explore ways in which to overcome the delayed effects of trauma.

Resources on Trauma

Herman, Judith. *Trauma and Recovery*. New York: Basic Books, 1992.

Middleton-Moz, Jane. *Children of Trauma*. Deerfield Beach, FL: Health Communications, 1989.

Terr, Lenore. *Too Scared to Cry*. New York: Harper & Row, 1990.

Common Connections

Growing up controlled can affect your relationships, career, emotions, spirituality, physical and mental health, and overall development. The following are common adulthood links to a controlled childhood. You might take note of which, if any, seem true for you.

Relationships

If you grew up with unrealistic expectations placed on you, you may expect too much or think too little of yourself.

If you had to act in a certain way for your parents, it may be hard to be yourself around others.

If your parents were judgmental, you may find yourself quick to judge others or hypersensitive to others' assessments of you.

If your parents acted unpredictably, you may be baffled by your feelings in relationships.

If your family was unsafe, all group activities may feel unsafe.

Love, Sex, and Intimacy

If you grew up in a tumultuous family, you may unwittingly be drawn to turmoil rather than health in your relationships.

If your parents devalued your rights, wants, and needs, you may also devalue your rights, wants, and needs in your relationships.

If your parents overcontrolled their relationships, you may tend to overcontrol yours.

If love meant control and intimacy meant danger, you may fear both love and intimacy.

If touch was misused or physical closeness connoted a painful vulnerability, you may have mixed feelings about touch and closeness.

If your parents modeled poor marital relations and erratic discipline, you may have trouble making commitments.

Parenthood

If your childhood hurt, you may equate all childhoods with hurt.

If your parents controlled you, you may have reservations about becoming a parent.

If you were raised with too much control, you may unwittingly react by raising your children with insufficient limits or control.

Professional Life

If your parents dictated your decisions, you may feel intimidated when making choices.

If your parents were unfair or abusive, you may see all authority figures as unfair or abusive.

If your parents misused their authority, you may find it hard to be an authority figure.

If your antennae were tuned to family turmoil, you may find yourself extrasensitive to workplace politics.

If your parents had unrealistic expectations and standards, you may not know when to quit or may give up too soon.

If your parents devalued you, you may lack confidence or underachieve at work.

Emotional Life

If you weren't allowed to express sadness, it may still be with you.

If tension was a way of life, you may still feel on guard.

If anger was forbidden, you may find it hard to let yourself get angry even when it is justified.

If you had to guard your emotions as a child, your primary feeling today may be numbness.

If your feelings were discounted or prohibited, you may be confused about what you are experiencing or how to express your emotions.

If "negative" feelings like anger, fear, or sadness were blocked, your access to spontaneous and positive emotions may also be blocked.

If your parents were depriving or using, you may automatically expect that when something good happens, something else you care about will be taken away.

Spiritual Life

If you grew up controlled, you may face great difficulty in finding a spiritual framework that fits your needs.

If your parents were punitive and controlling, you may view God as punitive and controlling.

Physical Health

If you were controlled, your body may carry the toll through stress-related illness, low energy, or chronic pain.

If you were deprived, you may seek relief through addictive or risk-taking behavior.

If your parents overcontrolled, you may face challenges from eating disorders.

If your parents exerted body control, you may have an unrealistic body image.

If your parents ridiculed or harshly punished you, you may find it hard to love and take care of your body.

Psychological Health

If your parents were perfectionists, you may still be trying to be perfect.

If your parents were harsh critics, you may have "inner tyrants" residing in your thoughts.

If your parents had double standards, you may give others the benefit of the doubt but blame yourself.

If your parents saw the world in black and white, you may have difficulty in seeing the grays.

If your parents regarded you as less than you are, you may also see yourself as less.

If your parents acted hatefully, you may feel unlovable.

If you rarely received praise as a child—or if whatever pride you did have was attacked—you may have trouble feeling appropriate self-esteem.

Overall Development

If your parents forbade dissent, you may have trouble asserting yourself.

If your parents controlled your every move, you may have difficulty in taking the initiative.

If your parents treated your independent spirit as sinful, you may seek continued dependence on your parents or other people.

If your parents infantilized you, you may feel you cannot assume adult responsibilities.

Each of these potential consequences is a common tendency, but they are not set in stone. The next chapter will show you several shortcuts to unhooking unhealthy connections between past control and present problems.

Sorting It Out

Since the information in the last few chapters has been emotionally loaded, take a few minutes to check in with yourself and notice any reactions you've had as you've read about the Dirty Dozen, Truth Abuse, controlling-family brainwashing, and the damaging consequences of family control.

For some of you, a lot of this may seem like old news. You may feel that you want to look forward and focus on how to achieve more peace and freedom in your current life and in the future. The rest of this book is aimed at helping you do just that.

For others, taking in this information may show your family in a new light. You may see your parents and your childhood in ways you never did before, which makes it impossible to return to your previous views and assumptions. While you may feel disoriented, or experience a sense of loss, take heart: You have undertaken a valuable, necessary step to health and maturity. Making childhood-adult connections can help you stop blaming yourself for things you never really did in the first place.

You may be feeling sadness. Sadness can be a hard feeling to sit with, but it's generally part of any transition from one set of values or identity to another. Making new connections between your past and present is part of making the transition to a more rounded view of your life.

You may feel angry. A saying in many self-help groups is, "The truth shall set you free, but first it'll make you mad." Anger is often misunderstood in society and particularly in controlling families. As difficult as it can sometimes be, anger, like all feelings, is best welcomed as a psychic road sign. **Anger is a valuable message from yourself that your boundaries have been violated or are in danger of being violated.** Feeling angry when your rights or boundaries are violated is essential to survival. Anger, in and of itself, is not destructive because you can express it in either destructive or constructive ways.

Finally, much or all of this part of the book may simply not dovetail with your experiences, values, and views. That's important to recognize and even more important to honor.

Exercise for Reckoning with the Costs of Growing Up Controlled

Envision being the physician or midwife who delivered you. As you hand your infant self to your parents, imagine looking your parents straight in the eye and telling them to always remember that:

- They will be the biggest single determinant in your development.

- You are different from them.

- You may be sensitive in ways they cannot understand.

Then envision handing them a box of bold warning labels, which are to be sewn on all your childhood clothes, carrying this message: WARNING: IF YOU RAISE YOUR CHILDREN WITH UNHEALTHY CONTROL, YOU WILL PLACE THEM AT RISK FOR DEPRESSION, ANXIETY, LOW SELF-ESTEEM, AND DISTORTED SELF-IMAGE.

How might your life have been different had your parents heard and heeded these warnings? Acknowledging wounds can bring grief, but remember: Your history is not your destiny. As an adult, you can provide for yourself and for those around you the nurturing, esteem, and healthy habits you were deprived of in childhood.

13

LETTING GO OF
THE LEGACIES
Giving up the Distortions of Your
Internalized Parents

And you shall know the truth, and the truth shall make you free.

—JOHN 8:32

Brainwashing, whether by a cult or a controlling family, is designed to hide responsibility and distort accountability—to keep anyone from daring to announce, "The emperor has no clothes."

Though brainwashing is powerful, it is not foolproof. If it were, everybody would be indoctrinated into cults, and no cult member would ever leave. Yet only a minority of people ever join a cult, and the majority of cult members eventually leave.

Similarly, parental overcontrol has its limits. Yes, you may sometimes feel as if you revert to being like a child when you are around one or both of your parents. Yes, you may sometimes feel as if you are in a trance, unable to assert yourself around someone who acts like one of your parents. But these feelings will pass. Parental control is not nearly as powerful now as when you were a child.

We explored in Chapter 11 that children have little choice but to internalize their parents' voices, especially the critical voices. For many who grew up controlled, the key to moving beyond blame and anger lies in finding new ways to cope with these critical inner messages. Internalized parents tend to perpetuate distorted views of the world. Recognizing these views as distortions is a powerful way to begin mastering the challenges in your life. Think of an obstacle or challenge in your life today, then see if any of the following five distortions are present in it:

1. Distortions of Power and Size

Distortions of power and size may have been fostered if one or both of your parents demanded absolute control, gave you little independence, or treated you in ways that made you feel small. As a result, you may automatically view yourself as less capable than others or, alternatively, as so big and powerful that you have to protect others from yourself. You may feel you need permission to do what is naturally your right. You may feel intimidated or, conversely, contemptuous in the presence of authority figures.

For example, Penny, the self-described "Good Girl-Miss Perfect" profiled in the previous chapter, had a distortedly small sense of her own power and size that left her viewing herself as small and flawed in relation to others. Alex, the burned-out super-salesman, had a distortedly large sense of his size and power that left him feeling that he was too big and too demanding for other people to cope with. Belinda, the thirty-year-old computer analyst, felt so small that she found herself cringing at work when she was around authority figures.

2. Distortions of Feeling and Wanting

Distortions of feeling and wanting may have been fostered if your emotions were banned, or inflated, or feared, or if your desires were shamed or thwarted. As a result, you may postpone attending to your feelings, thereby squelching them. You may view feelings like anger, fear, sadness, and joy as life-threatening and consequently overreact. You may misinterpret or be unable to listen to a loved one's strong feelings. Distortions of wanting may lead you to deprive yourself of legitimate yearnings or, alternatively, to live with unrealistic hopes. You may unconsciously come to expect pain in life and become uncomfortable when good things happen.

Penny, for example, found feelings so big, she felt as if she were going to break apart when others expressed strong emotions. Alex found feelings so frightening, he could not give himself permission to laugh or relax. Belinda found it hard to want to be a parent herself, automatically expecting that she would hurt her children as her mother had hurt her.

3. Distortions of Thinking

Distortions of thinking may have been fostered if truths were denied, perceptions discounted, parental responsibility avoided, and blame and shame heaped on you as a child. As a result, you may automatically preempt your attempts to stand up for your rights. You may accept unfair control from others, regarding it as normal. Distortions of thinking may lead you to avoid personal responsibility or assume responsibility for others' problems. You may also chronically doubt your perceptions or leap to conclusions based on all-or-nothing reasoning.

Penny, for instance, attended to life's "shoulds" and rules without questioning whether they were accurate or healthy for her. Alex couldn't let go of the "untidyness" of Friday afternoon business calls that wouldn't be resolved until Monday. Belinda feared making decisions because she was sure her choices would turn out to be wrong.

4. Distortions of Relating

Distortions of relating may have been fostered if closeness was elusive and dangerous, or if you were infantilized for too long or thrust into the role of caretaker too soon. As a result, you may feel unable to get close to others when you want to. You may trust people unwisely or, conversely, find that your ability to trust is stubbornly elusive. You may see others as threats, saviors, or objects instead of as human beings.

Penny, despite early warning signs, believed that her first two husbands would cherish her rather than treat her like "a thing." Alex felt trapped in his relationship dilemma of wanting "someone to get close to" but finding that when someone got close, he ran. Belinda could not bear to trust her boyfriend, though he had done nothing to betray her trust in their two years together.

5. Distortions of Self and Identity

Distortions of self may have been fostered if your intuition, initiative, pride, or needs were devalued. As a result, you may underrate your abilities, undercut your potential, or underplay your strengths. Distortions of self and identity may lead you to fear or banish parts of your personality, present a false front to others, or see yourself as an object instead of a person.

Penny found a positive self-image so elusive that she was incredulous when others complimented her. Alex, whose family had never asked his opinion, came to believe he had nothing interesting to offer others. Belinda found herself in the dark about her own abilities and desires after years of her mother's mixed messages.

How to Cope with Your Internalized Parents

Your internalized parents—earlier we also called them "inner critics"—traffic in distortions. Like viruses, they infected you in childhood.

You may never be able to entirely rid yourself of distorted thinking, but you can see it for what it is: the ghost of your parents' fears; a self-abnegating mechanism assembled within you before you had the wherewithal to prevent it.

Here are two tools that can help you abandon the distortions of your internalized parents.

Tool Number One: Revisit the Eight Styles of Controlling Parenting

Recall the eight styles of controlling parents from Part One (Smothering, Depriving, Perfectionistic, Cultlike, Chaotic, Using, Abusing, and Childlike). Ask yourself: "Do I treat myself the way my parents treated me?" Having identified your parents' styles, you may now have a big clue as to the nature of your harsher self-judgments. Of course, everybody has inner critics, those mental voices that carry on a lifelong monologue—judging, commenting, tempting, and threatening. But for those of us who grew up controlled, inner critics can mushroom into inner tyrants. Do you have an inner tyrant whose messages and criticisms are smothering, depriving, perfectionistic, cultlike, chaotic, using, abusing, or childlike? For example:

Smothering inner tyrant: *Who do you think you are?*

Depriving inner tyrant: *You don't deserve something that nice.*

Perfectionist inner tyrant: *Don't screw up.*

Cultlike inner tyrant: *You're doing it wrong and you're going to get in trouble.*

Chaotic inner tyrant: *You're losing control.*

Using inner tyrant: *You're selfish.*

Abusing inner tyrant: *You're worthless.*

Childlike inner tyrant: *You're incapable.*

Look at your self-criticisms. You might often feel that you are doing things wrong or that you can never perform well enough. Your thoughts, "It's wrong" and "I'm not good enough," represent perfectionistic themes. If you grew up with Perfectionistic parents, you may find that the overarching themes of perfectionism drive your thoughts and feelings even now.

The same is true with each of the styles of parenting. If you had Smothering parents, you may feel you cannot solve problems on your own. If you had Depriving parents, you may feel you must solve problems without others' help. If you had Cultlike parents, you may feel you have to know for sure before you can act. If you had Chaotic parents, you may feel too confused to easily solve problems. If you had Using parents, you may find yourself preoccupied with what effect your actions might have on others. If you had Abusing parents, you may be afraid to act boldly. If you had Childlike parents, you may worry that others are so fragile that your actions will harm them.

The themes of your inner tyrants—the "voice-over" your parents installed—tend to be loudest when you are acting or feeling counter to parental values or rules. When you acknowledge the distorted themes your inner tyrants are hawking, their actual self-critical messages become manageable. Rather than having to fend off each "It's wrong" or "I'm not good enough," you can instead see the bigger picture and note, "Ah, perfectionistic thinking, just like how I was raised." Children of Perfectionistic parents, for example, often feel flawed, not realizing that their parents found fault with everyone—even God. You might then remind yourself that you're in good company.

Your goal is not to deny your inner tyrants' existence or banish them from your psyche—that isn't possible. Your goal is to get to know your inner tyrants so you can puzzle out how they work. You can, in fact, set limits on your inner tyrants' influence without disowning their existence.

Once you have set healthier boundaries with your actual parents as well as with your inner tyrants, you may find you have a greater capacity to cultivate and listen to another voice within—the voice that speaks to you with care, compassion, and encouragement rather than judgment, bullying, and manipulation. This "inner nurturer" is derived from many sources: your deepest self and values; contact throughout life with nurturing others; and the "good parent" you may have dreamed of or caught glimpses of in your parents' finer moments.

Tool Number Two: Revisit the Dirty Dozen

While you had little choice about what your parents did to you, you have a great deal of choice about how to respond to your internalized parents/inner tyrants. Inner tyrants tend to use the same methods your parents used to control: the Dirty Dozen (direct control of food, body, boundaries, social life, decisions, feelings, speech and thought, along with indirect control through bullying, depriving, confusing, and manipulating). While appropriate self-control is a good thing, marching to the strident overcontrol of inner tyrants is emotional fascism.

Notice your next negative impulse or thought, then see if any of the Dirty Dozen are present. For example, do you: Harbor criticisms of your body and appearance? Give yourself little breathing room to make mistakes? Avoid new people and new experiences? Endlessly second-guess your decisions? Deny your feelings? Ignore your intuition? Scare yourself by frequently imagining worst-case scenarios? Deprive yourself by not asking for what you want? Think of yourself as stupid or flawed? Sometimes use food to punish or reward yourself?

Each of these critical messages derives from the Dirty Dozen. There is no need to be ashamed of such thoughts; we all have them from time to time. But you might pause for a moment in the midst of one to contemplate the thought and see if it is one of the Dirty Dozen. Next, ask yourself, "Is this critical thought about me really true, or is it really a fear or self-punishment?"

Recognizing the Dirty Dozen in your psyche can steer you toward shedding the distortions. For example:

If you live with great inner criticism and scant praise, you may want to block self-criticism and increase self-praise.

If you are compulsively driving yourself in reaction to parents who called you lazy, you may want to slow down. Conversely, if you are underperforming in reaction to or rebellion against pressuring parents, you may want to push yourself beyond your present comfort level in one or more areas.

If you don't pay attention to your emotions, you may want to gradually begin focusing on feelings. This action is like letting fresh air into a cut so it can heal: It hurts at first, but the healing is more complete. Something as simple as noticing at least five feelings a day can build awareness.

If your emotions feel stifled, you might allow more frequent emotional expression. One hallmark of growing up controlled is thinking, *Feelings shouldn't be*. But feelings don't have to be justified; they just are. Giving breathing room to your emotions can mean reminding yourself, "All my feelings belong. They don't need reasons to exist."

If you feel isolated from support and comfort, you may want to put more emphasis on seeking experiences that soothe and people who reassure.

If you feel deprived, you may want to gradually give more to yourself and allow yourself to receive more out of life.

If you have been giving your parents the benefit of the doubt to your own detriment, you may want to seek the guidance of a trusted friend or therapist as you question the validity of these viewpoints.

If you feel it's wrong to say no or point out what is wrong, you may want to spend more time saying no where appropriate or speaking out about what is wrong in your family, workplace, or society.

If you "left" your body years ago by numbing out, you might realize that your body never left you; spend extra time and attention caring for it.

Exercise for Letting Go of the Legacies:

Ask yourself: "Am I willing to have a better life than my parents?" Healthier parents want their children to have better lives than they did; controlling parents give children the message that they should not outdo the parents. But don't you deserve as good a life as you can have? While we can't completely control the quality of our lives, we can give ourselves permission to have a better marriage, career, emotional life, community of friends, and/or lifestyle than our parents. Giving yourself this permission can overcome lingering familial loyalties or injunctions that may unwittingly keep you from realizing your full potential.

Next: Why They Did It

The final chapters of Part Two will help you discover *why* your parents needed to exert such control over you. Doing so can help you make peace with your past.

14

WHY PARENTS OVERCONTROL

The more we idealize the past and refuse to acknowledge our childhood sufferings, the more we pass them on unconsciously to the next generation.

—ALICE MILLER

The roots of your parents' control can often lie in events that occurred long before you were born. As one woman I interviewed said, "Knowing my mother's past allows me to feel less like a victim. Knowing her past means she isn't just this person who did mean things to me. It's what happened to her, and her mom, and her mom's mom."

Controlling parents tend to be children of trauma. Trauma is an unexpected or shocking event or series of events that threatens your life or some vital part of yourself. This is not to say that everyone who has experienced trauma will grow up to be controlling. Nor is it true that every controlling parent experienced early trauma. But two factors are markedly common:

a. Controlling parents tend to have suffered severe and often multiple traumas as children.

b. They got little or no help in facing the consequences of the trauma.

This combination of trauma without help can lead to overcontrol. Four kinds of trauma are especially wounding to children:

1. **Loss of a parent**: Physical loss through death or abandonment; emotional loss through marked emotional rejection or neglect.

2. **Attacks**: Repeated assaults through physical or sexual abuse or malignantly negative parenting.

3. **Crises**: Societal or family crises such as war, natural disaster, poverty, severe illness, or the death of a sibling or close relative.

4. **Stress**: Chronic long-term stressors such as having an alcoholic or mentally ill parent, long-standing marital discord, or unhealthy family roles.

Among the more than eighty controlling parents of those I interviewed, the incidence of trauma was striking (see Notes on Research on pp 241–242 for the sources of the statistics):

- One in five children who later became a controlling parent had a parent die during their childhood—four times the national estimate of one in twenty.

- One in three children who later became a controlling parent had a parent who abused alcohol or was alcoholic—twice the national estimate of one in six.

- One in three children who later became a controlling parent had a parent who suffered serious depression or mental illness—more than twice the national estimate of one in seven.

- One in three children who later became a controlling parent was physically or sexually abused—nearly twice the national estimate of one in five.

- One in two children who later became a controlling parent grew up highly controlled—nearly seven times the national estimate of one in thirteen.

These little souls who grew up to be controlling parents faced stunning pain and loss. Many were children of a world war, the Depression, or the Holocaust. Several were cruelly cut out of their parents' hearts, homes, or pocketbooks. Many faced grave illnesses or life-threatening physical injuries. Some were simply children of bad luck and damaging happenstance.

The optimal recovery from trauma involves adequate time and a safe setting in which to talk about a traumatic event and the feelings that come with it. That's why many people intuitively want to talk about their accident or hospital operation in endless, repetitive detail, as if it is essential that they tell the whole story. It *is* essential because sharing can allow trauma survivors to relive the trauma at their own pace, which allows them to regain control after having had normal control of their lives taken from them.

Paths to recovery from trauma, however, are invariably missing among the childhoods of those who become controlling parents.

Well-meaning but misguided adults tell them to "Keep a stiff upper lip," depriving them of the chance to grieve over their losses. In socially isolated families, children lack anyone to talk with about their traumas, robbing them of the opportunity to receive compassion. In families that stress being perfect or avoiding showing weakness, traumatized children are blocked from working through the trauma. Among children with anxious or melancholy temperaments or who lack help in developing a sense of self-worth, trauma can take an even greater toll.

Underneath it all, trauma is about loss: loss of safety; loss of possessions; loss of love; loss of self. To a child, traumatic loss says:

People and things I need come and go unpredictably, and there is nothing I can do about it.

I can be attacked at any moment for reasons I don't understand, and there is nothing I can do about it.

Trauma victims who can't get help lose trust in a world in which people they care about are taken from them, in which people they depend on betray them, or in which disasters strike without warning. Controlling parents learn as children that it is risky to care about others because those they care about leave them or hurt them.

Lacking trust and expecting further losses, traumatized children's emotional growth may be stunted, leaving them with only a child's feelings and behaviors. As these children grow up, they may develop a philosophy of life that explains why they must control.

To an infant, the primary caregiver seems the source of everything: pleasure and pain, gratification and delay of gratification. As children mature, they require a parental balance of autonomy and closeness in order to learn that they can have an impact on the world and not be at the mercy of it. When a chief caregiver doesn't do a good job or traumatic circumstances interfere, children can grow up feeling that they will be annihilated by any form of abandonment or smothered by too much closeness.

Even as adults, our relations with others can be deeply influenced by what we didn't get, or got too much of, as children. We may tend to see the key people in our lives with the same distortions as our chief caregivers. If we felt deprived as children, our lives may be driven by striving for gratification from those around us. Alternately, we may give up and remain aloof because we are convinced we won't get any gratification at all. If we felt repeatedly threatened as infants, we may view those closest to us—spouse, children, boss, and/or friends—as potential annihilators.

Thus, parents who never felt *seen* as children may compel themselves to replay childhood dramas, insisting on being the center of attention, growing furious if admiration is not forthcoming. Parents who as children were smothered with malignant attention may find intimacy life-threatening. Parents who as children felt overridden may be terrified of being controlled as adults and, consequently, may adopt a stubborn, unreachable posture. They are likely, in any event, to experience difficulty in tolerating their own children's efforts to individuate.

Hellos and good-byes can be particularly difficult for some controlling parents. They may shake hands or hug awkwardly, if at all. Their first words when seeing you or others may be a sarcastic remark or a question that puts you on the defensive. Their demeanor may be strangely distant, happy-faced, or otherwise "out of synch" with the occasion—all reflecting their fear of losing their sense of self as others draw near or move away. Such controlling parents, gripped by the losses of their youth, may move into a world of projections and nonreality. They'll perceive others in a limited range of roles, such as threat, servant, or object. They'll remain constantly on guard and react to stress primarily with a child's emotional repertoire: tantrums, sulking, bullying, or selfishness. They'll do anything to avoid recognizing what healthier adults realize: *Life offers no guarantees of safety or happiness; wanting something doesn't necessarily mean you will get it; fearing something doesn't necessarily mean it will happen*—and they will live their lives and raise their children as if none of these realities exist.

Misguided Beliefs

Because they were traumatized and need to dominate, many controlling parents, consciously or unwittingly, tend to hold two beliefs:

1. I *own* my children.
2. My children *owe* me.

Of course, these beliefs aren't confined just to controlling parents. Many of us believe that children owe their parents respect, loyalty, and gratitude. Societies and religions foster these beliefs. Most of us were raised with them.

I offer a different view for your consideration:

Children owe their parents nothing simply because they are their parents.

Most of us love and respect our parents. It's natural to be thankful to the people who created and raised you, to be loyal to, approve of, and admire them. But while these feelings are natural, they are not debts. It's not a child's responsibility to pay in emotional currency for the right to exist.

Parents choose to be parents. Perhaps that choice is not thoroughly thought out or is made for the wrong reasons, but it's a choice nonetheless. Children, on the other hand, don't choose to be born.

Parenting, for all its hardships, has plenty of built-in rewards. Parents get to love deeply, receive love, express affection, and learn about themselves, life, and the world. Parents help a being come from nothingness to become a successful human. Parents can laugh, enjoy, play, revisit their childhood interests, and contribute to the world. Loving a child is its own reward. Being loved by a child is a special gift unlike any on earth. But with the choice of parenting comes the responsibility of learning how to raise children well.

All parents want to be appreciated by their children. Healthier parents recognize that appreciation is a gift their children *may* give, not something they *must* give. Healthier mothers and fathers may crave their children's love, respect, approval, and loyalty, but generally recognize that things like respect and approval must be earned by parents as well as by children.

Controlling parents, however, don't seem to know that truth. If they felt they had to earn their own parents' love, they may feel entitled to their children's love. In controlling families, need is stronger than love. Controlling parents need, expect, even command their children to love, appreciate, admire, listen to, and reflect well on them. Because controlling parents believe they own their children, they feel justified in such expectations.

Discovering Why Your Parents Controlled You

By better understanding your parents, you may achieve even more insight into their actions and character than they have about themselves. Granted, there is something inherently uncomfortable about having access to information our parents don't; it reverses the normal family roles in which parents know much that children do not. However, knowing more about your parents' pasts and deeply understanding them will benefit you in two ways:

1. It will allow you to deconstruct any larger-than-life or all-or-nothing beliefs and illusions you have unwittingly adopted about your parents or family.

2. It will help you actually see the roots of the distorted messages about yourself and others that your internalized parents send you.

As a family therapist, one of my first tasks is to generate hypotheses about why each family I see has the problems its members are seeking to fix. Part of my crucial and creative early work, then, is to gather enough information to generate these hypotheses, which often contradict each other. However, over time, one or more of these hypotheses will become especially compelling.

For example, when a family comes in with a child who is suddenly getting poor grades in school, I contemplate several factors. Did a recent change in the family upset the child? Are the parents getting along? Is the child getting clear limits along with positive attention? Is there drug or alcohol abuse, physical or sexual abuse in the home? Does the child suffer from a medical condition or learning disability? How are the parents reacting to the bad grades? How did the parents themselves do in school? How are siblings performing academically? Did something upsetting happen at school or among the child's circle of friends? Is the child or any family member depressed?

To explore these questions, I adopt a pose family therapists describe as "joining," in which I become a temporary member so I can see the family's world as they do. You may similarly find it useful to join your parents' world in order to generate hypotheses about why they controlled you. Of course, the idea of joining your parents' world may spark mixed feelings. You may want to know the reasons behind their confusing behavior. Yet after years of trying to make sense of your parents' actions or making allowances for them, you may feel ambivalent about continuing to do so. But remember, joining your parents' world is only temporary; it's a visit, not a merger. Visiting their world doesn't mean you condone their damaging control nor does it take away your right to feel angry, sad, or frustrated. But by joining, you may discover insights about your upbringing that can help you heal the wounds created there.

Seeing how your parents suffered as children may make you feel guilty for any feelings of anger about how you were controlled. This may not be rational, but it's a common feeling. Our parents *did* suffer. They were wounded as children. It wasn't their fault. Their unhealed wounds left them to parent in ways that wounded us. That wasn't their fault either.

At the same time, knowing that your parents suffered doesn't minimize what they did to you, how it made you feel, and what it may still be costing you. Just because your parents suffered doesn't mean your anger isn't justified. You may also feel compassion, sympathy, and grief for them even though they hurt you. By seeing how your parents were wounded as children, as well as seeing how they may have wounded you, you expand your ability to hold more than one "truth." Holding both truths and denying neither can be one of the most difficult dilemmas facing those who grew up controlled. It violates the all-or-nothing thinking of controlling families. It opens the door to questions and uncertainty. And it will make you stronger.

Controlling Parents' Early Trauma

Here are examples of how early trauma in the lives of the controlling parents of the people I interviewed may have translated into a controlling parental style. These admittedly simplified portraits list a parent's central trauma in childhood, her or his controlling style as a parent, and offer some hypotheses for *how* that trauma may have led to controlling behavior. Obviously, much more goes into personality development than a single trauma, and each parent's style is more complex than any thumbnail sketch. Still, I include these examples in the hope that they may help you develop your own hypotheses about why your parents may have needed to control you. From doing this, you'll gain a greater understanding of their actions and also better recognize and understand your internalized parents' ongoing negative messages.

Henry: Smothering Parent

Henry's Trauma: Death of a Father

Remember Sally, whose Smothering father greeted his daughter's coming out as a lesbian by ripping the pink triangle bumper sticker off her car and asking for years afterward when she was going to find a man and get married? When Sally's father, Henry, was four, his father died in a farming accident, so the boy was raised in an all-female home by his mother, grandmother, and aunt. In adulthood, Henry became a Smothering parent who micromanaged Sally's eating and bedtime habits.

Hypotheses

1. Henry may have grown up with unrealistic notions of his power and his duty to control everything. Henry's daughter Sally speculates, "As he was the only male in the home, I suspect they told my father, in a well-meaning way, 'You're the man of the house.' But my father somehow took it all in as, 'I am responsible for everything.'"

2. Perhaps Henry thought that if he kept everything in control, he could ward off future family tragedies.

3. Perhaps he was afraid that he, too, might die young and wanted to do all he could for his children while he was still around.

4. Perhaps, given that Henry was raised by a houseful of females, he was angry with women in general and transferred that anger onto his daughter.

5. Perhaps he never had a chance to develop his own sexual identity, and the idea of his daughter's homosexuality intimidated him.

6. Perhaps Henry was simply doing what he thought was best for his children in the hope of sparing them the pain he had suffered.

Nathan: Perfectionistic Parent

Nathan's Trauma: A Brother's Death

Remember Will, whose Perfectionistic father rode herd on him before swim meets but rarely praised his son's victories? Will's father, Nathan, was five weeks old when his older brother died, yet Nathan was not told of it and only learned of the loss years later when a relative brought it up. In time, Nathan became a Perfectionistic parent, rarely becoming emotional about anything, yet able to talk for hours in minute detail about the technical challenges of his work as an engineer. He raised his children to always be in control of their feelings.

Hypotheses

1. Even when a child is not directly aware of a family member's death, grief can hover, and unspoken grief becomes all the more terrifying by virtue of its mystery. Though Nathan almost certainly had no conscious memory of his brother, he was born into grief

that was never spoken about and grew up in a harsh, controlled world of secrets. Some people who are terrified of feelings such as grief habitually intellectualize in order not to be emotionally overwhelmed.

2. Perhaps Nathan focused on technical details because they made sense, in contrast to an early loss and cover-up that made no sense.

3. Perhaps, even as a child, Nathan sensed that life was fragile and tried to be perfect so that he, too, wouldn't die.

4. Perhaps Nathan's early family cover-up induced in him such a suppression of emotionality that he never learned how to deal with feelings—his *or* others.'

5. Perhaps Nathan, aware that he'd lived when another being hadn't, was still trying to justify his existence by doing everything "by the book."

Rita: Depriving, Perfectionistic Parent

Rita's Trauma: Emotional Abandonment

Rita was treated "like a princess" by her mother until she was five. Then a sister was born, and Rita was suddenly dropped from favor as her mother transferred her affection to the newborn. Two years later, Rita's mother did the same thing to Rita's sister when a third girl arrived. Rita grew up thinking her mother hated her, and consequently became a Depriving, Perfectionistic parent, living an emotionally barren life based on the philosophy, "You can't trust people." Rita micromanaged her daughter's eating and dress habits and rarely praised her accomplishments. Her daughter never recalls seeing Rita cry.

Hypotheses

1. Perhaps Rita's Depriving philosophy of "You can't trust people" stems from her loss of legitimacy in the eyes of the most important person in her life—her mother. She must have grown up wondering: *Did I really deserve the princess treatment? Did my mother really mean all the nice things she said about me? If so, why did she stop saying them?*

2. The sudden loss of her mother's affection was a striking deprivation: How could Rita hope for positive things in life when she'd

been unable to trust something as basic as a mother's love? Like many who lose parental favor, she came to expect that life would continue to reject her.

3. Perhaps Rita could not provide constant love for her daughter because she had never known constant love in her own childhood.

4. Perhaps she was too depressed to see her daughter's needs.

5. Perhaps any show of independence by Rita's daughter made her feel that the girl was no longer part of her, which rekindled early feelings of abandonment.

6. Perhaps she could see her daughter only as an object, since she had herself felt cast off like one.

Larry: Chaotic Parent

Larry's Trauma: Manipulation and Disinheritance

In his teens, Larry worked three jobs to earn money for college, but his father—who insisted his son follow in his footsteps and join the military—appropriated the money and gave it to Larry's sister for college. With little financial support, Larry acceded to his father's wishes and joined the army, but he never forgave his father and refused to visit him on his deathbed. His father, in turn, left Larry a one-dollar inheritance, transferring his wealth to his grandchildren, who today are fighting over a substantial fortune. As a father, Larry had a volatile, Chaotic presence and always seemed on a vendetta against somebody. He repeatedly disowned his children when they disagreed with him.

Hypotheses

1. Larry was no doubt enraged by his father's manipulations and may have felt compelled to take his rage out on the world. Perhaps he was jealous and resentful of his children because they ended up with "his" inheritance.

2. Perhaps he felt so afraid of being controlled that he tried to dominate everyone around him so *they* couldn't control him.

3. Perhaps he never healthily separated from his parents, so when it came time to help his children emotionally separate, he didn't know how; disowning was his only model.

4. Disowning is a final act of control. Some battering spouses kill their mates because, if they cannot have their spouses under their control, they don't want anybody else to. Similarly, rather than acknowledge the reality that he had lost control of his children, perhaps Larry symbolically killed them by disowning them.

Helen: Using, Abusing Parent

Helen's Trauma: Assault and Banishment

Remember Ellen, who was coerced every night to tell her mother, Helen, how beautiful she was? In Europe during World War I, seven-year-old Helen was raped by a "friendly" soldier, then beaten by her parents for being raped. When she was thirteen, her father died, and she was sent to live with relatives, though her mother kept Helen's younger sister with her for reasons Helen never knew. Helen was physically abused by her relatives and never saw her mother again. In time she became a Using, Abusing parent, hypersensitive and depriving, coercing her daughter into a nightly ritual of doing Helen's hair and nails and hitting her if she made a mistake.

Hypotheses

1. As a parent, Helen seemed fixated on telling stories about the servants and household elegance of her childhood. "She glamorizes her past and idolizes her mother even though her mother abandoned her," says her daughter Ellen. By idealizing her mother and her past, perhaps Helen skirted the horrible memories of her childhood abuse and powerlessness.

2. Perhaps Helen never integrated the recognition that others can be both nurturing and rejecting, so she could see her mother as only all good or all bad. Unfortunately, Helen also tended to regard her daughter as all bad and treated her accordingly.

3. After the trauma of rape, disfavor, and abandonment, perhaps Helen felt imperfect and unclean and, even years later, needed her nightly beauty ritual as compensation.

4. Perhaps after the horrible deprivations of her childhood, she simply wasn't able to nurture others.

5. Perhaps her abusive control of her own daughter was a way in

which Helen expressed rage over her own abuse and abandonment.

6. After being treated as a child with little value, perhaps Helen could not see her daughter as having any value beyond that of serving her.

George and Paula: Depriving, Perfectionistic Parents

George's and Paula's Traumas: Death of Their Mothers

You may remember David, whose Depriving, Perfectionistic parents didn't let him have film for his Brownie camera because they were trying to discourage his artistic interests. As David told me about his bleak childhood, I began wondering if his parents grew up feeling deprived and joyless, which was exactly how they raised David. It turns out that both David's father, George, and mother, Paula, lost their mothers at an early age.

George lost his mother before age five, at which point George's father sent him off to boarding school. As George grew up, his distant father made him work in the family business, paying him the minimum wage. As an adult, George rarely talked about his feelings or his childhood. He developed hardening of the arteries and died at forty-six.

Paula lost her mother when she was four. Following her mother's death, her father sent Paula to live with another relative and moved away. Paula was raised by an aunt who treated her like an intruder; she didn't see her father again until she was fourteen. When the aunt's husband died in World War II, the aunt developed phobias and depression. As a parent, Paula became a "severe worrier" who obsessed about order.

Hypotheses

1. When a parent dies, children are especially vulnerable to others' treatment of them. Deprived of their mothers, George and Paula grew up virtual prisoner-orphans, George at boarding school and Paula, who felt like an intruder, with her unstable aunt. Lacking the nurturing a mother can provide, both George and Paula grew up with something warm and vital missing from their lives. As parents, though they may have done their best, they couldn't help but raise David with that same emptiness.

2. Perhaps they feared David might die, just as their mothers had done, and couldn't bear to attach emotionally to him.

3. Perhaps they obsessed about order and routine in order to distract themselves from deep depression.

4. Perhaps George, who was given so little by his father, didn't know how to give more to his son.

5. Perhaps Paula's years of watching her depressed and phobic aunt convinced her that the world was unsafe, thereby leading her to be an anxious, rigid mother.

Mike: Using Parent

Mike's Trauma: Injury and Abuse

Remember Magda, whose immature, Using father, Mike, bought her birthday gifts, only to play with them himself? When Mike was six, his mother gave him a yard of rope from the family store to make a jump rope. When his father discovered Mike jumping rope, he cut the rope into one-inch pieces, threw them into his son's face, then beat him. At age nine, Mike suffered brain damage from a car accident but apparently received little or no treatment or neuropsychological testing. As an adult, Mike, in charge of quality control at a large corporation, became a Using parent who frequently beat his children.

Hypotheses

1. Mike probably got emotionally stuck very young. When children become emotionally stuck, immense grief and anger get locked up inside them. Some controlling parents never move past that stage of dramatic, childlike feelings, never learning how to have perspective on their emotions. They only know how to act out their feelings as a young child would.

2. Themes of violence and scarcity are reflected in Mike's childhood. Since abuse and control were all he knew, perhaps all he could do as a parent was follow that model.

3. Perhaps Mike had a deep need for power over others. What more appropriate job for someone powerless as a child than to oversee everyone else's work as the head of quality control?

4. Mike saw violence modeled and may have grown up full of rage at his deprivation. Perhaps his abuse of his children served as revenge against his father.

5. Perhaps his head injury played a physiological role in his violent behavior and poor impulse control as an adult.

6. Perhaps he needed to be the center of attention as an adult because he had felt so deprived as a child.

Lucy: Childlike Parent

Lucy's Trauma: Ostracism

After Lucy's father skipped town when she was five, her mother, because of finances, felt she had to quickly remarry, eventually marrying a man with whom she had four children. Lucy, the oldest child, felt like the black sheep of the family because she was the only sibling with a different father. When, at thirteen, she was caught stealing a neighbor's milk, her mother sent her to reform school, claiming she'd get an education, and more food than she would get at home. Her mother also told Lucy that she was a "thief" and deserved to be sent away. Lucy in time became a frail, Childlike parent, repeatedly sacrificing her daughter Molly to her husband's wrath. Lucy lived under her husband's thumb, doing errands and making phone calls only at the times he permitted. She was also overinvolved in her children's bodily functions, giving enemas, douches, and medications.

Hypotheses

1. Lucy must have grown up feeling illegitimate in many ways. Her father had vanished; she was not only an outsider in her new family, but was also labeled a "thief," then banished. Perhaps Lucy thought, being not legitimate, that she had no rights and deserved nothing good. Perhaps she covered her feelings of illegitimacy by seeking a stronger spouse who would make decisions for her, even if it entailed the abuse of her children.

2. Perhaps she sought a stronger spouse as a father figure to replace the father who had left her. She may have been too terrified to question her husband's abusive behavior for fear she would lose him also.

3. Given how much pain Lucy faced early in life, perhaps she assumed that suffering was just a part of growing up.

4. Perhaps she was depressed and overwhelmed and simply couldn't cope with parental responsibilities.

5. Perhaps she never overcame the rejections of her youth and remained an emotionally abandoned little girl, looking for others to care for her in ways she'd never experienced as a child.

6. Perhaps Lucy overcompensated for her mother's lack of interest in her by becoming intensely interested in her children, right down to their bodily functions.

7. Perhaps she felt ostracized because she felt somehow bad or flawed. Her obsessive attention to her children's cleanliness may have been a misguided attempt to fix or cleanse herself.

8. Perhaps Lucy felt anger at her mother but could never express it directly; her children thus became unwilling substitute targets for her anger.

Lloyd: Perfectionistic Parent

Lloyd's Trauma: Death of a Mother

Remember Chip, whose adoptive father, Lloyd, "dropped him" when he was found to have learning disabilities? Lloyd's mother died giving birth to him after doctors had warned her that childbirth could be life-threatening. Saddled with the legacy that his mother had died so he could be born, Lloyd was raised by a domineering, stoic father and became an aloof, Perfectionistic father himself.

Hypotheses

1. How could Lloyd forget that his mother had died so he could be born? As a result, he may have felt guilty, bad, or cursed. Perhaps seeing his son's "flaws" reminded him of his deepest fears about himself. Being close to his "flawed" son may have been too hard for Lloyd to tolerate.

2. Perhaps he felt at fault for his son's learning disabilities just as he felt at fault for his mother's death. He may have pulled away from his son rather than confront self-blame.

3. Lloyd's life began with a catastrophe and proceeded with little warmth. Perhaps he was so afraid of losing another loved one that he couldn't allow himself to get close to anybody in case they, too, left him.

4. Having grown up with little warmth from a manipulative father, Lloyd may not have known how to attach emotionally to another human being, especially a male.

Compassion

I can't help but feel compassion for these anguished children who became controlling parents. Their little souls suffered stunning pain and loss. As misguided as their adult overcontrol became, it reflected deficits in their own childhoods.

Of course, it's too easy to assume that traumatized children will inevitably become controlling parents. In fact, some controlling parents of the men and women I interviewed had no apparent notable childhood traumas. In addition, no parent is affected every minute of his or her life by childhood trauma. Traumatic effects wax and wane; they may be evident in some parts of a parent's life and virtually invisible in others. While a parent may be a controlling tyrant in the family, he or she may show little overcontrol at work or at play, in some situations or at certain times. The Perfectionistic, Cultlike, or Smothering parent may be a model worker. The Abusive or Using parent may be a fierce competitor in sports or business. The Childlike or Depriving parent may be a brilliant researcher or bookkeeper. The Chaotic parent may be an artistic genius.

Furthermore, most parents, no matter how controlling, have moments of caring, grace, and a desire to see their children happy. Controlling parents may want to stop their children's suffering, along with their own, but just don't know how.

The long-term effects of trauma tend to be most prominent when people are stressed, in new situations, or in situations that remind them of the circumstances of their traumas. Unfortunately, being a parent is all three: stressful, new, and almost always the trigger for memories of their own childhood traumas. An intimate relationship like parenting is a fertile arena for control because intimate relationships are the settings for most traumas. In intimate relationships, we are most vulnerable, and vulnerability is unwelcome to children of trauma.

Exercise for Understanding Your Parents' Roots

Think about what you know about your parents' early lives. Did they face the loss of a parent, an attack, a family crisis, or long-term stress as listed earlier in this chapter (see page 130)? If so, did they get help? How might their traumas have colored their world views and affected their emotional lives? You might develop at least three possible hypotheses, then see how each fits. (The section "50 Reasons People Control in Unhealthy Ways" in the next chapter may help spark your hypothesizing.)

If you don't know much about your parents' early lives, you might want to get them talking about them. This can be a win-win situation: Your parents can validate their early existence and you can gain insight into your family's roots. In addition, if your parents tend to focus attention on you in negative ways, asking them to tell their stories allows you to step into observing mode, shift the focus away from yourself, and feel less reactive to what your parents might say and do. Of course, you may not want to ask about your parents' roots if you feel they'd become abusive during such conversations. If that's the case, relatives can be extremely helpful in filling in the blanks about your parents' early lives.

15

CONTROLLERS' FEARS

It is when power is wedded to chronic fear that it becomes formidable.

—ERIC HOFFER

Trauma engenders fear—a key commonality among controlling parents. Knowing the needs and fears your controlling parents carry within them, you'll begin to understand why they controlled as they do or did. You'll also be better able to grasp the source of the negative messages from your internalized parents. Both your actual parents' behavior and your internalized parents' critical messages, no matter how mystifying, are driven by five fears:

1. Fear of being seen as flawed

2. Fear of feeling powerless

3. Fear of feeling invalidated

4. Fear of feeling vulnerable

5. Fear of losing emotional control

One of the fascinating aspects of human behavior is that it often compensates in reverse. Someone who feels particularly small may strut around acting larger than life. Someone who feels adrift in an emotional rapids may become a stoic. Someone who fears rejection may reject others first.

In the case of controlling parents, these defensive actions become maladaptive. Feeling flawed, controlling parents pretend they are perfect. Feeling small, they act big. Feeling afraid, they frighten others.

Feeling bad about themselves, they shame others. Feeling wrong, they insist on being right. Feeling doubt, they confuse. Feeling deprived, they withhold.

Controlling parents compensate in ways that cost both child and parent dearly. The need to feel powerful and worthy becomes a life-or-death crisis for such parents. Avoiding vulnerability is suddenly a matter of survival. Why? Because powerlessness, vulnerability, and unworthiness remind them of their desperate childhood days, when they felt flawed, full of doubt, helpless, out of control, and afraid for their lives. Controlling parents (and, for that matter, your internalized parents) will do anything to avoid recognitions we all must face:

There are forces and people more powerful than I am.

There are people who don't need me or fear me.

Time, death, and illness will humble me; this is the price of being human.

Rather than face these realizations, many controlling parents chose childlike coping behaviors: denial, tantrums, bullying, running away, and/or playing take away. They become, as Elan Golomb wrote, "psychologically hard of hearing" (152). Unconsciously, they adopt myths about themselves: the self-made man, the perfect mom, the good provider, the in-control dad, the biggest son of a bitch in the jungle. These myths give parents the illusion that they are in total control of their destinies, masters of the universe after a childhood of feeling little mastery. To admit anything different would once more leave controlling parents feeling powerless. This may explain why some of them seem disconnected from the present, often unaware of their surroundings and feelings. Living in the moment risks loss of control and lacks guarantees—exactly how controlling parents felt as children.

Controlling parents are often unaware of why they act as they do. If they realized what lay underneath their maladaptive behavior, they'd have to face their painful childhoods, their dependency on others for their feelings of self-worth, and their desperate hunger for the symbols of success. They'd have to face the fact that they are as controlled as anyone else.

Controlling parents rarely learned as children that facing their feelings or admitting their limits can be healing. Because they try to control everything, they tend to think that others, including their children, are doing the same. Since most controllers want to be sure they are never dominated, they move to control others first.

In short, being a controlling parent is a defensive action. A combination of factors—how the controlling parent was raised, lack of knowing better, external events, internal needs, and the footprints of trauma—leave controlling parents, unless they get help, playing out a lifelong defensive drama. Even as adults no longer at the mercy of childhood trauma, most controlling parents dare not acknowledge how powerless they once felt. They may even deny that the trauma occurred. They may fear that exploring their memories will make them reexperience feelings as real and frightening as they were when the trauma occurred.

There is a certain logic to this behavior—the distorted reasoning of traumatized children whose sense of self splintered at an early age. Controlling parents are wounded children whose life was warped by dramas they didn't create.

Because as children they didn't get sufficient help, attention, and love, controlling parents generally feel that they are not adequate—though they may act in quite the opposite way. As adults, they may seek assurances of self-worth through watertight rules, beliefs, and practices. Their overcontrol is a futile effort to secure guarantees that they will be loved and safe rather than powerless, invalidated, or out of control. Yet it is costly because:

- Parents who fear being judged as flawed can never let others see them as they truly are.

- Parents who need to feel powerful must always be on guard against threats to their power.

- Parents who fear invalidation cannot tolerate questions or uncertainty.

- Parents who fear vulnerability view everything and everyone as potentially threatening.

- Parents who must avoid feeling out of control are likely to miss out on joy, spontaneity, and love.

Parental Control Is Not Personal

Because they were frightened, your parents may have taken personally much of what happened in their lives. You don't have to. If you take nothing else from this chapter, I hope you come away with the realization that your parents' control wasn't personal. They didn't dominate you because you were bad, inadequate, did something wrong, or were cursed by God. The reasons had to do with them, not you.

Ironically, no matter how domineering they are, most controlling parents think they don't dominate enough; when you're terrified of the world, you can never do enough to protect yourself. A further irony is that, while many of us spent thousands of hours trying to figure out how to make our parents more accepting and less controlling, they would have controlled no matter what we did. There's no way you could have stopped them.

Overcontrol isn't personal, it's generational—until someone breaks the cycle, it's a white-knuckle response to trauma. You can be that someone. Part Three of this book will show you how.

50 Reasons People Control in Unhealthy Ways

While control is necessary for living, dysfunctional control is not. We can never know absolutely why others act as they do, but we can make educated guesses that lead to greater understanding and compassion for parents, other people, and ourselves. Here are fifty reasons why people control in unhealthy ways. Notice any reasons that strike a chord in explaining why either your actual parents or your internalized parents control.

Then, you might review the list and ask yourself if any of these reasons explain why you sometimes control yourself or others in unhealthy ways.

Cognitive Reasons: People Overcontrol Because They Believe

1. Others will take advantage of them.
2. Total control of others is possible and that they are controlling for others' good.
3. The world is unsafe and control can ward off danger.
4. Disagreements can destroy people and being criticized is life-threatening.
5. Values and lifestyles differing from theirs are wrong.
6. They are superior to other people.
7. Situations are a zero-sum game in which there is always a winner and a loser.

Intergenerational Reasons: People Overcontrol Because They

1. Were raised with excess control and did not fully or healthily separate from their parents.

2. Grew up feeling abandoned or smothered and came to see others as potential abandoners or smotherers.

3. Felt overridden and deprived as children and are terrified of being overridden and deprived as adults.

4. Never felt *seen* as children and now insist on being the center of attention.

5. Had misguided models of how to treat people.

Emotional Reasons: People Overcontrol Because They

1. Fear their needs for safety and dependency and consequently intellectualize instead of facing their feelings.

2. Attempt to avoid a huge reservoir of grief, anger, or regret and see others' emotions as unsettling reminders.

3. Are convinced they won't get gratification so they remain aloof.

4. Have poor body images or conflicts about sexuality and are jealous of others' healthier or younger bodies.

5. Possess poor emotional coping skills and cannot teach others how to deal with feelings.

6. Envy others' good fortune.

7. Are depressed, anxious, addicted, and/or have poor impulse control.

Power/Gratification Reasons: People Overcontrol Because They

1. Feed off the energy of others.

2. Personalize others' actions.

3. Need to feel grandiose because they are petrified of feeling weak or powerless.

4. Are addicted to control, which, like a drug or a drink, brings on a rush.

5. Need to channel their desire for revenge and to feel others are dependent on them.

6. Are just plain mean.

Unconscious/Existential Reasons: People Overcontrol Because They

1. Are unwilling to admit they have weaknesses or fears.

2. Are angry with themselves, a spouse, a boss, or their own parents, but displace their anger onto others who are not as threatening.

3. Act in ways that mirror their fears (e.g., a father who grew up in a chaotic home may be obsessed with order; a mother who was treated as stupid in childhood may exhort her children to be "smart").

4. Are in denial about their control and the pain they cause others.

5. See others as the cause of their problems and are overly suspicious about what others are doing.

6. Become inflictors of pain to avoid feeling like helpless, passive victims.

7. Disown weak aspects of their selves and can't tolerate anything small, helpless, and weak around them—like children.

8. Fear accepting the humbling reality that few people like, but most of us eventually accept: *We all have some power, but events are often dangerous and random and there will always be things outside our control.*

9. Need coercive rules and rigid beliefs to maintain the status quo and tidy up life's messy questions.

10. Need to outlaw dissent to prevent anyone from pointing out that perhaps they are not as perfect or as in control as they would like to believe.

11. Are trying to distract themselves from their own problems, flaws, or feelings.

12. Resent others' hobbies or close relationships because they feel their influence shrinks and fear that others will come to love someone or something else more than they love them.

Self-Esteem Reasons: People Overcontrol Because They

1. Feel they cannot stand up for themselves and don't deserve anything any better than what they have.

2. Need others' approval and want to be seen as perfect.

3. Hope another's accomplishments will accord them status, as in "My son the doctor."

4. Cannot cope with the demands of parenthood or adulthood.

5. Excessively value beauty, fame, power, or money.

Interpersonal Reasons: People Overcontrol Because They

1. Assign people to a limited range of roles, such as those of servants, masters, or objects, and respond to them as such.

2. Never integrated the realization that others can be both nurturing and rejecting, so they keep others at a "safe" distance.

3. Felt like "things" around their parents, so view other people as objects.

4. Are inept at distinguishing their own needs or fears from those of others.

5. See others' bodies as extensions of themselves (one man said his Perfectionistic father saw him as a "walking, talking, stuffed animal sprung from his loins").

6. Feel defeated in reaching their goals or regret not following their dreams and want to save others from making the same mistake.

Circumstantial/Societal Reasons: People Overcontrol Because They

1. Are overwhelmed by their unmet needs and/or face financial, social, work, physical, or marital crises.

2. Do not make healthy relationships a priority or subscribe to societal and cultural values that foster overcontrol.

Exercise for Discovering Why People Control

Think of an incident in which someone tried to control you in unhealthy ways. Which of the above reasons might best describe why they acted as they did? How does understanding why someone else controls affect how you feel about them and about control in general?

Part Two—A Summary

It's hard to acknowledge that you had little choice and control in your early life. It may feel demeaning to admit that you're still struggling with problems spawned by your parents decades ago. Yet acknowledging your lack of choice and control in early life can spark a freeing recognition:

Given your upbringing, many of your problems make sense—and they are not your fault.

Yes, your problems are yours to solve. But they do not reflect innate character flaws, the lack of the ability to love or be loved, or a lack of competence. Many of your adult-life problems may stem from early situations which you had no power to alter. Blaming yourself can only hurt. Healing is not about blaming yourself or others. Healing involves:

1. Seeing the controlling-family brainwashing in your past.

2. Seeing the "trances" that are induced by the internalized parents in your present.

3. Seeing when you are controlling yourself or those around you just as your parents controlled you.

4. Learning to appropriately trust rather than automatically control.

The final section of this book will explore many of the healing strengths you possess, including ones you may not even realize you have.

PART THREE
Solving the Problem

Power can be taken, but not given. The process of the taking is empowerment in itself.

—GLORIA STEINEM

In the Introduction I wrote:

1. *You aren't responsible for what your parents did to you, they are.*

2. *You are responsible for what you do with your life now, your parents aren't.*

The remainder of this book is about what you do with your life from now on.

Solving the problem of growing up controlled has three steps:

<u>Step One:</u> *Emotionally leaving home* by separating from the hurtful aspects of your upbringing, parents, and family role.

<u>Step Two:</u> *Bringing balance* to your relationship with your parents.

<u>Step Three:</u> *Redefining* your life.

Emotional healing is like physical healing. If you cut your finger, you clean the wound and protect it from infection with a bandage. If you break your leg, you set the bone and wear a cast to protect it from further trauma. This allows your body's natural healing process to work.

It's the same with emotional healing. When you're emotionally wounded by a controlling childhood, "cleaning" the wound means facing your true past and speaking about it. And the "bandage" or "cast" that protects these wounds from further injury is *emotionally leaving home*. This doesn't necessarily mean a physical separation from your parents, but it may entail letting go of counterproductive links with them and your upbringing.

You cannot mend a broken bone faster by telling it to "heal more quickly." Healing a broken leg means wearing a cast, which can make walking difficult. Similarly, emotional healing may mean changes in habits that at first feel awkward.

Like physical healing, emotional healing can happen twenty-four hours a day without conscious effort. You may not know exactly how a cut heals; you just notice that each day it gets a little better. Similarly, people who begin emotionally separating from a controlled upbringing frequently notice over time that they develop more positive values and a greater sense of freedom, often without knowing precisely how.

Emotional separation opens the way for you to *bring balance* to your relationship with your parents, whether they are living or dead. Emotional separation also permits you to *redefine* your life and yourself in terms of who you really are and where you really want to go, not in terms of your parents or your past.

Step One: Emotionally Leaving Home

16

SEPARATING FROM UNHEALTHY FAMILY TIES

Honor thy father and thy mother.

—EXODUS 20:12

Why isn't there a commandment to "honor thy children" or at least to "not abuse thy children?"

—BEVERLY ENGEL

Emotionally leaving home means emotionally separating from any hurtful or counterproductive links with your parents, your past, or your family role. In so doing, you discover new aspects of yourself and recover nuances of your personality long obscured by a controlled upbringing.

Family therapist Murray Bowen wrote that the level of emotional separation among family members, which he termed "differentiation," is one of the key determinants of emotional health (409). As Bowen pointed out, you had little choice about how differentiated your family was. Yet you have great choice in your adult years as to whether you attain a greater degree of differentiation and emotional separation than your parents did.

Emotional separation is comprised of three elements:

1. Observing

2. Declaring your independence

3. Mastering the challenges of separation

In healthier families, emotional separation takes place naturally and gradually, accelerating as children reach their teens. In controlling families, this emotional separation, if it is allowed at all, takes place unevenly. Controlling parents often lack the skills to help their children separate; young adults from controlling families often leave home with overwhelming emotional baggage because a natural separation couldn't take place.

Emotionally separating from your parents is a major step in adult development. You may not know how big a blot a controlled upbringing is on your life until you begin to differentiate. Those I interviewed described how energizing it was to gain perspective about their painful pasts and take the reins of their lives. Many described feeling as if the "fog" had lifted. Others testified that, by just stepping back, they began to focus more on their own needs and less on their parents.'

Remember, you grew up under controlling-family brainwashing and it can take time and effort to free yourself. Remember, too, that there are many paths and a range of paces to emotional separation—there is no single right way or best speed. Often separation starts slowly, with baby steps over the course of many months. It can be enhanced by cultivating an internal ally—the strong part of you that is always there, watching—as well as by relying on external allies such as supportive friends, family members, therapists, teachers or mentors, self-help groups, or literature.

As a child, you may have tried to distance yourself from the pain of being controlled by complying, rebelling, distracting, dissociating, or outdoing your parents. While these methods helped you survive, they kept you emotionally tied to your parents and may still do so today.

Do one or more of these five coping strategies describe a posture you still find yourself taking with your parents?

> *Complying:* You seek the path of least resistance, doing what your parents want or what you think they would want, even when it means forgoing your own best interests.

> *Rebelling:* You seek the path of greatest resistance, automatically opposing your parents to avoid being controlled, even when rebelling harms you.

> *Distracting:* You make light of or change the subject, rather than directly face your parents' control even though it reduces your stature in your own or others' eyes.

> *Dissociating:* You daydream, sleep, seek out an addiction, or virtually avoid your parents rather than face the threat of parental control even though it costs you awareness and aliveness.

> *Outdoing:* You compulsively try to control yourself or your parents when you are around them, even though it heightens your stress.

As an adult, you have more options than complying, rebelling, distracting, dissociating, or outdoing. A valuable first step is observing.

Observing

Emotional separation begins with stepping back and observing. This means observing your parents, the effects they have on you, and your responses to them. Despite the "do something" orientation of many controlling families, often the best thing to do initially in a troubled relationship is simply to observe.

Observe how you feel before, during, and after contact with your parents. As one woman put it, "When I go home, I have to play the good daughter and be polite. I become part of this machine. My body is there but not my heart and spirit."

After contact with your parents, do you feel valued or devalued? Content or irritable? Trusted or betrayed? Optimistic or hopeless? Accepted or judged? Confident or flustered? Energized or fatigued? On a scale of one to ten—ten being the biggest—what size do you feel in relationship to your parents? How does that compare to how you want to feel?

Ask yourself whether you have contact with your parents because you want to, or out of obligation, as Ellen Bass and Laura Davis suggested in *The Courage to Heal.* Your answer can tell you volumes about what the relationship both provides and costs you.

In addition to observing how you act and feel around your parents, notice how your parents act around you. A turning point in my relationship with my father came some years ago when I went to his house for my nephew's christening. I wasn't looking forward to being around my father because our relationship had grown icy and tense, but I did want to be there for the christening. A colleague suggested I use the visit to observe how my father controlled.

During my two-day visit I took my colleague's advice and mentally counted each controlling thing my father said or did. The total was in the dozens: He dictated when we ate, where we ate, when we got together, when we could approach him, when we had to halt other conversations to listen to him, even what we touched in his house. I was nearly forty and my father's control, as pervasive as it was, was infinitely milder than it had been when I was a child. Yet I still found myself hesitating and wondering what I had done wrong. I was receiving only a taste of what had been my steady emotional diet, but it was enough to dispel any doubts about how my father's control had haunted me as a boy.

By emotionally detaching myself in order to observe, I didn't have to fight anything my father did; everything he did was data I could learn from. Through observing, I realized his fundamental operating principle: Avoid domination by others. This helped me to see why he had raised me as he did and why he continues to control. Perhaps most important, I could appreciate how his fears and operating principle might prohibit our having the kind of relationship I wanted.

Contact with parents can become bearable, even a growth opportunity, by your becoming an information gatherer—a sort of family anthropologist. Observe how your parents control and what sets them off. Observe how you feel around them and whether your behavior changes in their presence. You may notice how like children controlling parents can be; in many ways, they have children's resources, views, and emotions. Their tantrums are children's tantrums; their attempts at control are children's attempts.

Observing means embarking on a research project. Rather than a "decision project" in which you have to know all the answers so you can act, a research project involves a transitional period of time in which you don't have to have instant answers. Everything that happens provides information that helps you to better meet your wants and needs. You don't have to be in a hurry; in fact, observing is best done when you don't feel pressured.

Of course, observing your parents won't work if the contact is too destructive or costly, but the benefit of observational experiments is that nothing can really go wrong. Everything that happens is data for you to synthesize in your own way, in your own time, for your own purposes.

Exercises for Observing

1. **Count the control moves.** In your next conversation or visit with your parents, count each controlling thing they say or do. You might notice which of the Dirty Dozen methods they use (control of food, body, boundaries, social life, decisions, speech, emotions, and thoughts, along with bullying, depriving, confusing, and manipulating). Notice what precedes or seems to trigger their efforts to control. Notice how their control makes you feel. Your method of counting can be silent or overt. Keep a mental tally or take notes. Use a golf-score clicker or cough each time they do something controlling. Don't tell them what you are counting.

Then ask yourself what it must have been like for a child to grow up around such control.

2. **Imagine that you are a *60 Minutes* reporter building a story about your parents.** Use a tough-minded *60 Minutes* reporter's eye to see their apparent motivations, quirks, and inconsistencies.

3. **Visualize your parents in an imaginary shrunken terrarium that duplicates their living room.** Imagine you are wearing a "Far Side" cartoon-style white lab coat as you observe their habits. Replay key incidents from your childhood, and study how your terrarium-dwelling parents controlled. When you stand at a distance, you'll see a lot about their styles.

Declaring Independence

Emotionally leaving home means declaring independence, but declaring independence in and of itself doesn't make you free. The American revolutionaries found that out in 1776 when they declared themselves free but had to fight to prove it. In the Declaration of Independence the American revolutionaries said to the world:

a. All people have inalienable rights.

b. The British misused their power and violated American rights.

c. That after unsuccessfully trying to work it out with the British, the Americans felt their only choice was to sever relations and declare themselves free.

It is a beautiful, ground-breaking document for human rights. Declaring their willingness to have the world judge the truth, the revolutionaries asked for nothing, threatened nothing, and declared that they needed nothing.

Emotionally separating from controlling parents may feel like a revolution: life-threatening, yet exhilarating. It's a bit like learning to walk. As a toddler you were so intent on walking that if you fell or bumped into things or looked awkward, you didn't care, because it felt so good to join the world of upright creatures and steer under your own power. Emotionally separating is like this; you may fall, bump into obstacles, or feel awkward, but autonomy brings joy and, oh, the places you can go!

Declaring your desire for independence, even silently, can empower you. It breaks the trance induced by growing up controlled and initiates the powerful process of separation. Declaring independence allows you to see how different you are from the others with whom you identified. It can pave the way to stunning new perspectives about yourself and the world.

Declaring independence from your parents is different from denying their emotional impact on you. They are, after all, your connection to hundreds of ancestors, each of whom had a part in who you are today. You'll always have parents; their voices are present in your psyche even after they die. In denying your parents' role in shaping you, you risk denying a part of yourself. In acknowledging their role, you reclaim part of your heritage.

Exercises for Declaring Independence

1. **List the ways in which you have already achieved independence.** Note the ways in which you have broken from parental practice and dogma—how you are fundamentally different from your parents.

2. **Write your own Declaration of Independence.** Read the 1776 Declaration of Independence, list your rights (the "Bill of Rights for Those Who Grew Up Controlled" on page 239 may help stimulate your thinking), then consider how one or both of your parents violated them, what those violations cost you, and what you have tried to do about it. For many, declaring independence is an internal, private step never discussed with parents. For others, it may include some communication with one or both of them. (Chapter 18 will explore the issue of confronting parents who control.) You may even want to send your Declaration of Independence to your parents. If you do, sign it boldly, like John Hancock. Remember: You are doing this for you, not for them. That's what independence means.

3. **Independence Day.** Take your birthday or other significant day as your personal Independence Day holiday. It might be the anniversary of the day you said no to controlling parents, literally left home, or turned a corner in emotional separation. Celebrate your Independence Day as one of your most special holidays.

4. **Channeling Dad/Mom.** If you find yourself acting as your parents did, lightheartedly acknowledge that you are temporarily "channeling" Dad or Mom. Such actions are, in fact, like being possessed by a ghost from the past: Now, when I act bullheaded like my father, my friends chime in, "He's channeling Al."

Another way to externalize your parents' influence is to see yourself as having a "Dad Attack" or "Mom Attack," since the experience is like being overtaken with a coughing or a sneezing fit. Be assured that it will pass.

Mastering the Challenges of Separating

Emotional separation nearly always begins with pain. During the course of healing the pain may intensify, just as a cut or bruise hurts more in the early stages of mending.

Emotional separation accelerates when you give voice to the recognition that your parents controlled you in unhealthy ways. This can be hard to acknowledge because, if you grew up controlled, disagreeing with your parents may have had severe consequences. As a child, when you realized that something was wrong in your family you may have felt helpless to do anything about it, and grew frightened when you perceived your parents' limitations. This is a distressing recognition. After all, if the people you measured yourself against for so many years acted unhealthily, what does that say about the reliability of the lessons they taught you.

Yes, separation does bring losses, but these are necessary losses. You may have to give up the hopes or fantasies that your parents will protect you or be your "best friend." But controlling parents probably judge you and love conditionally, so protection and closeness aren't realistic goals anyway.

As you emotionally leave home, you may feel a conflict between remaining "true" to your parents or being "true" to yourself and risking losing parental love. You may temporarily become an emotional orphan. This takes great courage. It is one of the deepest hurts possible to admit that one or both of your parents were not there for you, didn't see you, and didn't protect you. As time goes on, the pain tends to recede, though it may never entirely vanish. Remember, you lived for years in an atmosphere of brainwashing. It takes time and work to free yourself.

Willingly or not, you have invested much of your lifetime in your family, so one of the hardest parts of separating is letting go of your identity as a family member. It can be challenging to differentiate between the ingrained warnings and criticisms of your internalized parents and the helpful messages of your true voice. (Hint: Internalized parents tend to say "Don't," "You can't," "You should," "You

shouldn't.") You may find that part of the grief comes from losing values with which you identified. Giving up earlier "versions" of yourself,
while healthy, can bring sadness.

Yet, unlike when you were an infant, your parents are no longer
the most important persons in your life. They are not crucial to your
survival as an adult, and your relationship with your parents is only
one of many aspects of your life today.

It may be hard not to include your parents in your process of emotional separation. It's natural to want parental approval and blessings
for any important challenge you undertake. But you don't need it and,
by definition, you can't have it this time. There's great value in keeping your separation separate from your parents. They don't need to
know how or why you are separating, what's hard about it, or what
feels validating about it. Save those reports for your friends, mate,
and/or therapist.

It may cause you deep pain to see how much you tried to earn
parental approval, how unresponsive your parents may have been, and
how you continued to try anyway. Yet grieving over the losses of your
childhood is central to healing from growing up controlled. As we saw
in Part Two, many controlling parents suffered extreme losses early in
life but never grieved over them. As a result, they tried to control a
world in which they felt utterly out of control. By facing your grief,
you reduce your own need to overcontrol.

Separation also includes taking stock of how you are similar to as
well as different from your parents; of how you were influenced by
them along with how you shunned their influence. The first time
someone told me I sounded and acted like my father, I was shocked.
Most of my close friends who have met my father saw this instantly. I
was the one who didn't. I tried to be different from my father because
I wanted no connection to his control. But I am connected. I can deny
it, but it would be a lie. I can avoid it, but it is there. I can fight it, but
I would only be fighting part of myself.

Discovering some of your parents' unhealthy traits in yourself is a
phase of healing. One man I interviewed confessed that he secretly
hoped misfortune would befall his friends who had happy marriages
because he wasn't in a romantic relationship himself. It didn't take him
long to realize that he was following the pattern of his Using father's
jealousy. "Instead of trying to deny my envy," he told me, "I simply see
that it's there. Seeing that it's a part of me begins the healing."

Have compassion for yourself. Remind yourself that abusive control did happen and that it was not your fault.

Feelings Accompanying Separation

Because of who your parents are—the people who gave life to you—it is natural to have intense and conflicting feelings as you separate. Any of the following emotions and concerns are common during emotional separation from a controlling family:

Feelings: grief; disillusionment; shame; numbness; self-blame; guilt; disloyalty; anxiety; vulnerability; fear of abandonment;. anger; depression; sadness; exhilaration; disorientation.

Yearnings: wishing your difficult feelings would go away; desire to recapture a real or imagined sense of having a close-knit family; sadness, mourning, and envy when seeing healthier families or friends' close relationships with their parents; desire for revenge or compensation from your parents.

Worries: fear you'll be angry forever; fear you are becoming controlling like your parents; fear you are selfish, unfeeling, or lacking in compassion; worry that your parents may disown you; worry that your independence will hurt your parents; worry that your parents may die with bad feelings left among you; fear of retribution from your parents.

Sensitivities: indignation over past and present control; increased sensitivity to others' comments about you; anger over what parental control has cost you; anger if a parent who abused you denies that the abuse happened; anger at having emotionally fed your parents for so long; embarrassment at having been controlled; sadness at molding parts of your life to meet unhealthy parental demands; pain at realizing how petty some of your parents' actions may have been; change in eating or sleep habits; spontaneous crying.

Conflicts: pressure for certainty accompanied by enhanced self-doubt; feelings of freedom, followed by a disorienting sense of "Now what?"; satisfaction if your parents face hard times, followed by guilt over such feelings; impatience with your pace of change, followed by worry that you are going too far or too fast; a feeling of having a dark part of yourself that you want to get rid of; confusion about why your parents did what they did; anger taken out on yourself or others; feeling you should forgive your parents but finding it difficult; anger when someone tells you to forgive your parents.

As distressing as these concerns can be, make them your allies. Each emotion, after all, provides you with information about some part of you because your hope, frustration, and fear speak to you through your feelings. Don't be surprised if strong feelings surface. Chances are, strong feelings were stuffed down in your childhood and are still there to be expressed. You might not like some of the initial feelings you experience—anger, sadness, or grief, for example—but these are a testament to the fact that your emotional separation is working. You are doing the unfinished work of a teenager leaving home. The difficult feelings will pass.

Since feelings just happen, there's no such thing as a "wrong" emotion. While we can control how we *express* our feelings, we can't control *what* we feel. If you grew up controlled, you grew up with "shoulds" about emotions. Now is the time to let go of these shoulds.

It can be difficult to reconcile our understanding with our feelings. For example, can we be angry with our parents for hurting us, yet also know that they were in pain most of their lives? Can we acknowledge that we didn't deserve the way our parents treated us, but also realize that our parents didn't deserve how they were treated as children? Can we accept that even though our parents were not at fault for what happened to them as children, they were responsible for what they did to us? Can we feel both anger and compassion for those who cost us a great deal yet gave us a great deal? Can we reconcile the realization that our parents may not have loved us consistently because no one loved them consistently?

To all the above questions, the answer is yes. Though it may seem contradictory, **healing from growing up controlled comes, in part, from cultivating contradictions rather than avoiding them.** Our parents said that feelings have to make sense or that if our feelings and thoughts were in conflict, one or the other had to be wrong. But feelings are not logical and aren't designed to match our thoughts. By cultivating contradictions, you see that there is often more than one "truth" in human relations.

The feelings that accompany separation emerge when they are ready to. They may erupt spontaneously or they may simmer. Remember, healing is a transition, and transitional periods bring unfamiliar emotions and new behaviors. During transitions, it's important to give yourself time to grieve over what you are leaving behind, to explore your feelings, and to try new approaches to life. It's also important to acknowledge what you can control and what you can't. In transitions, as in living, we often have little control over our feelings or the twists and turns ahead. But we can fully choose our goals and actions. That freedom of choice makes transitions both frightening and exciting.

Exercises for Emotionally Leaving Home

1. **List the advantages and disadvantages of emotionally separating.** In one column, list all that emotionally leaving home may cost you; in the second column, write down all it may gain you. There are losses as well as gains in leaving home. When you have a clear view of the potential losses and gains, you can best choose your path.

2. **Cultivate contradiction.** We want life to make sense and we like to feel certain. But life doesn't always make sense, and there are so many truths. Take five minutes and identify at least five contradictions, equally true but opposing statements. You might particularly cultivate contradictions that carry an emotional charge, since these provide the biggest opportunity for growth. For example:

   My parents did their best and yet they hurt me.

   I know it's not good for me and yet I want it.

   I love him/her and yet I hurt him/her.

   I tend to resist new experiences even though they would be positive and enjoyable.

   I know others aren't perfect, but sometimes I expect them to be.

   The strength of this exercise comes not in resolving the contradictions but in expanding your ability to hold more than one truth at a time. Despite the either-or, all-or-nothing habits of controlling families, both sides of many contradictions are equally true. Often the healthiest form of resolution is just to let both sides be without seeking a conclusion that denies either one.

3. **Give yourself an ideal send-off.** Ideal parents would have debriefed you as you left home by saying something like:

   "I know I may have called you lots of bad things and treated you badly at times. I know you have suffered. At times I had my reasons, but at other times my behavior just reflected my weaknesses. Now that you're leaving, I want to apologize. I want you to leave home unencumbered. Forget all those negative things I said about you. You are none of them. All those times I told you 'Do,' 'Don't,' 'Should,' or 'Shouldn't'—feel free to ignore them. You're an adult and it's your life. I trust you to make the right choices."

While there are no ideal parents, and while your parents probably didn't say anything like this, you can symbolically provide yourself with the ideal send-off you never received by writing it in a journal, role-playing it, visualizing it, or meditating on it.

Step Two: Bringing Balance to Your Relationship with Your Parents

Why do you hasten to remove anything which hurts your eye, while if something affects your soul you postpone the cure until next year?

—HORACE

Emotionally leaving home gives you the breathing room to create a healthier balance in your relationship with your parents. A healthier balance, in turn, pays dividends in your relationships with mates, close friends, and family members.

You grew up in an out-of-balance family. In seeking balance as an adult, you may face one or more of the following dilemmas:

1. How can I set healthier boundaries with my parents?

2. Should I confront my parents?

3. Can I forgive my parents?

4. Can I accept my parents?

5. Should I reduce or break contact with my parents?

The next chapters show various ways in which those who grew up controlled have faced these dilemmas.

17

HOW CAN I SET HEALTHIER BOUNDARIES WITH MY PARENTS?

Self-defense is nature's eldest law.

—JOHN DRYDEN

If the pendulum swung too far in your family—if you weren't allowed to feel your feelings, speak your mind, or go your own way—it's time to make the pendulum swing back. Yet relating to controlling parents poses many challenges.

What if a parent continues to intrude or abuse? How do you cope with your conflicting feelings? How often should you visit, call, or send a gift or card? What if, despite your best efforts at separating, a parent continues to meddle in your social life, career, or child raising? How can you maintain individuality without either freezing your parents out or forfeiting your independence?

The commonality in these challenges is setting boundaries. Remember the Dirty Dozen—direct parental control of your eating, appearance, activities, social life, decisions, speech, feelings, and thoughts, and indirect control through bullying, depriving, confusing, and manipulating? Each of the Dirty Dozen was a boundary violation, which your parents may still be committing today. Even if your parents are dead or you have little contact with them, your internalized parents can still bully, confuse, deprive, or manipulate you.

As an adult, you can undo the Dirty Dozen by establishing the boundaries that as a child you couldn't set. Often simply acknowledging a boundary violation leads you to act in your own best interests.

For example, as a child you may have been forced to answer parents' intrusive questions. Now simply declining to answer inappropriate questions sets adult boundaries in the place of the childhood ones that were ignored at will.

If your parents hurried you, you now have the right to use time as your ally, telling them you need a few days to mull over a parental request, invitation, or comment.

If your parents physically crowded you or invaded your privacy, you can now set the limits of your personal space.

If your parents meddled in your social life, it's up to you to choose what personal activities you'll share with them.

If your parents gave you gifts with emotional strings attached, declining their gifts or openly addressing the "strings" can help you achieve an equal footing.

If you grew up with few allies, bring a supportive partner along on your next visit to your parents—or make sure you can speak with a friend or therapist by phone if necessary. If your parents live some distance away, staying somewhere other than at their house can help balance out past privacy violations. Maintaining your normal routine of diet, sleep, exercise, entertainment, and personal growth practices during parental visits keeps you grounded in your adult sense of self, balancing your parents' control over your food, body, and activities.

When you were a controlled child, your parents chose their amount of access to you. As an adult, you may find that altering that access brings a welcome change in perspective. For example, if contact with your parents is painful, you may want, for a time, to erect a protective boundary. This is done **externally** by reducing or temporarily halting contact. Or it's done **internally** by emotionally detaching, thereby having less at stake in what your parents do, say, or think about you.

A trial period or "vacation" of limited or no contact with one or both parents doesn't mean you're forever cutting contact. You have the right to claim separate time or space in any adult relationship when it's in your best interests. Setting limits on your parents' access to you may mean simply going through the motions of sending occasional cards or making perfunctory phone calls. This can be a temporary trial period or it can be a long-term strategy built on acceptance of the limitations of the relationship. Turning points such as standing up to your parents or saying no for the first time are rites of passage in separating from and gaining balance in controlling families. They boost your self-esteem.

Your goal is finding peace from a painful childhood and freedom from past and present control. To achieve peace and freedom, you choose "which bridges to cross and which bridges to burn," as one woman I interviewed put it. There is no predetermined end point, no "right" or "best" kind of relationship to establish with controlling parents. If you feel pressured to make the "right" choice, you may be under the spell of perfectionistic thinking: Even now, the parents in your head are trying to control how you relate to the parents in your life.

It's hard to view one's parents neutrally. Yet for many who grew up controlled, one indicator of successfully separating and balancing comes in being able to see a parent as "just another person." That means applying the same standards and values, no more and no less, to parents as you do to friends or associates. This doesn't mean you won't feel loyalties and conflicts. Rather, it means that after a lifetime of elevating or degrading your parents—of seeing them as larger than life or smaller than insects—you're coming to see them as people, less than perfect, just like you.

This equalization may come from letting go of needing, or expecting, or hoping for anything from your parents, including your hopes about what parents "should" have provided that you didn't get and maybe never will. This takes time and may bring grief, but it also offers freedom. As you give up emotional attachment to what your parents could, should, or do offer you, you may find that you are left with a relationship with simply your parent the man or woman. It's easier to set healthy boundaries with, accept, and have compassion for another from whom you need nothing.

Despite how massively you may have been controlled, keep in mind what I call the **Ten-to-One Principle of Healing:** *Any action done by choice, consciously and deliberately, undoes the effect of at least ten such actions done unconsciously or in which you had no choice.*

Each time you deliberately act in your own best self-interests by countering controlling thoughts, people, or situations, you undo more and more of the effects of past control.

Stories of Boundary Setting: Undoing the Dirty Dozen

Here are stories about how some of those I interviewed set healthier boundaries with their parents by counteracting Dirty Dozen–style boundary violations.

A Christmas "No": Elizabeth

Most controlled children were rarely allowed to say no to their parents. For many, a key to setting adult boundaries is declining to follow family rituals. Remember Elizabeth, the thirty-one-year-old travel agent whose Perfectionistic, Using mother endlessly rehearsed her daughter in how to say hello when answering the phone? A turning point in Elizabeth's relationship with her mother came one Christmas. Elizabeth, a student in the middle of a painful relationship breakup, final exams, physical ills, and money worries, told her mother she was not planning to visit for Christmas. Her mother threw a fit and wrote to her daughter, "I hate you and I'm never going to think of you anymore."

Recalls Elizabeth, "She was like a kid who'd lost her toy." Yet being rejected by her mother after a lifetime of deprivation and manipulation ultimately helped Elizabeth grow: "When she cut me off, I felt, 'Oh my god, this is what I've always dreaded.' But I also realized that she could no longer get at me. Her letter validated that I had every right to feel angry. I could see how terrifying it would be for a child to receive that kind of rage from her."

By saying no to the visit, Elizabeth took a big step toward balancing a childhood of forced attendance on her mother, which eventually allowed Elizabeth to resume the contact on an equal footing.

Balancing Emotion Control: Sharon

As controlled children, our feelings were outlawed, warped, or denied. When feelings conflicted or didn't make sense, we were usually told those feelings were wrong. Balancing can mean welcoming all your feelings surrounding your parents.

Sharon, a thirty-one-year-old graduate student, has vowed to speak her truth whenever she has contact with her Smothering father, a Holocaust survivor. Paradoxically, doing so has given her more freedom to feel compassion for him. "My father is just like a little boy who is hurt. I try to see beyond things and know that what he is doing, he does to protect himself. Sometimes I just want to tell him I forgive him and let him back in. But he would only hurt me again."

For Sharon, balancing has meant embracing her conflicting feelings: anger; love; compassion for her father's limitations; disappointment over her family's failings; and her desire to live her own life. Rather than trying to reconcile these emotions by placing them in a tidy package—which was the way her family approached emotions—Sharon has found vitality through facing all her feelings, however untidy the process.

Finding Her Own Answers: Ina

Remember Ina, the fifty-three-year-old social worker whose Chaotic mother ordered her to be smart and pretty, with top grades and lots of dates, yet did everything she could to discourage her daughter from acting smart or feeling pretty around the house? Despite her mom's mixed messages, Ina made it a point to develop her own philosophy of life. At age eleven, she was reading philosophy books from the library. In college, she studied philosophy, psychology, sociology, and literature.

After college, Ina embarked on a plan for freedom as ambitious as that of the eleven-year-old Ina's quest for her own belief system. She wrote stories, kept a journal, joined a psychodrama group, wrote her autobiography, meditated, danced, and studied martial arts. At age twenty-eight, she moved two thousand miles from her parents, seldom visited, and made and wrote only "empty and ritualistic" calls and letters to them. She focused on "more fulfilling, more promising, and less damaging" relationships than the ones with which she had grown up. Collecting ancestral family stories and "poring over them like an archeologist" was helpful in unearthing the roots of her parents' behavior.

For Ina, seeking her own answers has helped balance her family's thought control.

Balancing Boundary and Food Control: Robin

Robin, a fifty-three-year-old design artist, recalled a recent phone conversation with her Using, Depriving mother during which her call-waiting signal beeped. "After I returned," she says, "my mother tried to wheedle out of me for five minutes who had called."

When Robin was a controlled child, she couldn't have secrets from her mother; now she could state firmly that it was none of her mother's business: "I told her I'd hang up if she kept asking me. She did, and I did."

Recently while dining at her mother's, Robin asked her not to put a rich sauce on the chicken. "She did it anyway. I threw the entire dish out. She was annoyed, but the more you give, the more she wants."

For Robin, blocking her mother's intrusions into her privacy, social life, and eating habits has helped her find empowerment by doing what she could never do as a child.

"Acted Like I Didn't Have Parents": Claire

When Claire, now a thirty-six-year-old real estate agent, was in her twenties, she reduced contact with her Smothering, Abusing parents to virtually nothing for several years: "I acted like I didn't have parents."

Then came a two-year period in which she sent them an occasional card. "I didn't want to totally write them off but I felt resentful about giving them gifts on holidays," she admits. "By sending only a card I was honoring my feeling of not wanting to give gifts." Over time, her feelings changed, and she sought more contact: "Eventually I made gifts so beautiful that I wanted to keep them for myself, but I sent them to my parents."

For Claire, allowing her feelings to evolve at their own pace helped balance a childhood of being told what to feel.

Emotional Distance: Brittany

Remember Brittany, the twenty-three-year-old sales representative whose Chaotic, Abusing, alcoholic mother used to hit her, banish her to her room, then show up all smiles with a gourmet dinner minutes later? Brittany wrote a letter confronting her mother about the older woman's continuing abuse and manipulation. Brittany then took three months off without any contact. "I told her that she was both an angel and a devil to me and that I couldn't take care of her anymore," Brittany recalls. "I felt free for the first time after that."

Taking a break despite her mother's threats was a big step in balancing the manipulation with which Brittany was raised. While her mother didn't change very much, Brittany found the three-month break an eye-opener. Later, studying communication theory in college helped Brittany identify her mother's baffling patterns of communication and helped balance the confusion of Brittany's childhood.

"Made Me Furious": Julia

Twenty-six-year-old Julia, a clerical worker, visited her Depriving, Chaotic mother and Abusing stepfather for Christmas. The turning point came when her stepfather "was in this incredibly bad mood, having a midlife crisis," she remembers. "I was afraid he might fly into a rage and hit me. I told him I was afraid of him. After I left, my mother told me that he was hurt and felt abandoned and that I should write to him. That made me furious. Since then I've been dropped by him and have had no contact."

After Julia's mother entered therapy and eventually apologized for allowing the abuse, Julia felt better: "But it's not that simple. I can't wipe out the way I feel about myself. In a curious way, the feeling of being betrayed helped me realize how little I still get from my mother and stepfather, and that helped me to move on."

For Julia, balancing has meant standing her ground as she sorts through a mix of feelings: anger over not being protected by her mother; the desire to be close to her; and grief at what growing up controlled has cost her.

Mother's Day Break: Tina

Remember Tina, the forty-eight-year-old social worker whose Smothering mother made her wear a "Please Do Not Feed Me" sign in public as a girl? Tina's break from her mother's grasp came, appropriately, on Mother's Day. She recalls, "I looked at every Mother's Day card in the store and began crying because those lovely messages were not in my experience. I couldn't buy a card that said, 'You were such a wonderful mother. You were always there.' So I left the store. I thought about taking something to my mother. I looked at the phone for hours. Finally I realized it was eight o'clock, and I'd done it. I hadn't given her anything.

"I was terrified. The next day at work I typed her a note saying, 'I can't see you anymore. I'm in therapy and feeling too angry to see you.' My mother called my sister, crying and screaming, 'How can she say this? I have been such a good mother.' She called me every name in the book. It became forbidden to mention my name in her house."

For three years Tina and her mother had little contact. However, with time and therapy, Tina began to heal. "Each day got easier," she admits. Three Mother's Days later, Tina sent her mother a "little bitty" Mother's Day card. Her mother responded with a card. After a few more months and warmer cards, Tina asked to visit her: "My mom sent back a big bunny card and said 'Of course you're welcome to come.'"

When Tina arrived, her mother threw open the door and said, "We will not talk about the past, right?" According to Tina, "She was still making the rules. I did not agree, and in the years since I have told her a lot about how she hurt and abused me."

For Tina, balancing meant taking the step she had always dreaded by breaking contact with her mother, then speaking her truth after reestablishing contact.

Potential Risks of Boundary Setting:

- May spark bad feelings in your parents or family
- Can lead to feeling guilt, grief, or disloyalty
- May lead to retaliation from your parents

Potential Benefits of Boundary Setting:

- Can offer protection from further control
- Can provide a "breather" in which to gain perspective
- Can provide empowerment by balancing childhood boundary violations

18

SHOULD I CONFRONT
MY PARENTS?

*Sometimes it's better to oppose and be angry, especially when a view of
yourself has been imposed on you.*

—SHARON, 31, A GRADUATE STUDENT

Children form their identities, in large part, by expressing themselves
and asserting their wills. Yet growing up controlled means having
your speech, feelings, and thoughts stifled. That's why a controlled
childhood hampers development.

Confronting parents by speaking up about past or ongoing control
helps some people balance a childhood of speech control. After not
being allowed to speak out for years, many people feel a compelling
need to confront their parents by letter, tape, phone, or in person.

Others, however, may not want to directly confront their parents
but instead find healing through symbolic confrontations such as: writing a letter to your parents that you never send; visualizing an imaginary conversation with a parent; or telling supportive friends what you
would like to say to one or both parents. For example, one woman
whose Abusing father terrorized her found healing in talking with
female friends about her traumatic upbringing: "It was more helpful
than formal therapy."

Some points about speaking out and confronting:

Confronting Is Entirely Your Choice

Confronting is a choice, made by you in your own best interests. It
is not necessary to say a word to your parents about how they hurt
you.

As Susan Forward suggests in *Toxic Parents*, confronting parents is not designed to retaliate, punish, or get something positive back. Rather, confrontation is designed to overcome the fear of facing your parents, speak the truth, and determine the type of relationship you can have from now on (235).

The purpose of confronting is to voice your truth. Once you have done that, consider the confrontation a success, no matter how your parents respond. "Changes in [parental] behavior are not the measure of whether you are making progress," wrote psychiatrist Harold Bloomfield in *Making Peace with Your Parents* (112).

Weigh the Risks and Benefits

Add up the possible risks and benefits in advance of confronting a parent. Think about how your mother's and father's reactions might affect you. Ask yourself what you want, what you expect, what you fear, and what, if anything, you need from them. Be realistic in your expectations. Few people like to be confronted, and your parents may ridicule you, dispute everything you say, retaliate, display no reaction, or simply not recall their unhealthy behavior. After all, if they had a history of hearing and respecting you, you probably wouldn't be confronting them in the first place.

It's true that the potential risks of confronting include a more contentious relationship, parental retribution, or, in some cases, the loss of active contact with a parent. But the potential benefits include greater individuation, self-esteem, and peace whether or not relations with your parents actually improve.

Losing an active relationship with a parent can be a big price to pay. Yet allowing continued parental control and abuse, being stranded on a mountain of feelings, and accepting treatment from a parent you wouldn't tolerate in any other relationship is costly as well. Only you can assess the potential costs and benefits and make your choice, realizing that all choices have risks.

Plan Your Message

Confrontation is a tool for healing, not a goal in itself, suggests therapist Mike Lew in *Victims No Longer*. You can confront in any way you want and change your mind about it at any time. However, many people who grew up controlled want at least three things from their parents:

1. An acknowledgment of the unhealthy control and its costs

2. An apology

3. Some effort by parents to reduce or cease the overcontrol

You may not get any of these, at least in the form you want, but it's important to realize what you're looking for. If your parents become abusive, end the confrontation. This doesn't mean you have failed; when confronted, controlling parents frequently respond with the exact behavior you're pointing out.

You may never achieve a true dialogue with your parents. As controllers, they probably have little interest in your views, particularly if they're negative. It's maddening when controlling parents deny responsibility for their actions. "Don't look at me," they say. But who else could have had the impact on you that they did?

It's helpful to remind yourself of some basic truths before and after a confrontation:

Your parents used unhealthy control.

Their control hurt you and cost you.

You have the right to voice all your feelings about being controlled.

You aren't responsible for what your parents did to you, they are.

You are responsible for what you do with your life now, they aren't.

These truths sometimes get obscured during a confrontation, so it's useful to seek additional sources of clarity. Have a supportive person standing by to "debrief" you after the confrontation. Susan Forward's *Toxic Parents* is a helpful source as well. If you choose to confront your parents, the book offers a detailed discussion of just how to do so.

Speaking out, regardless of the outcome, balances the distortions of so many years. People even find that, in confronting, their parental saber-rattling giants have faded into mild and friendly ghosts.

Speaking up Is a Process, Not a One-Time Event

As I've said, confronting has its own timing. Just as there are risks in rushing to forgive, so can there be risks in rushing to confront. You may want to confront your parents on your terms or turf to avoid replicating family control.

Once you initially express yourself, new feelings or insights surface. Again, it's up to you whether or not to express them. The benefits of a confrontation can take time to surface. Your parents may react positively and then backslide, or react negatively and then improve. You can't make them listen if they don't want to. Sometimes, when parents get sick, fear death, face losses, get lonely, or remarry, their denial does tend to break a bit, which allows them to hear you. Saying what's most dreaded can kick off a transformation. If, as a result, your parents show a willingness to understand your pain and acknowledge their part in it, you may have the beginnings of a healthier relationship. If they show little desire to change, you may want to temporarily detach from them emotionally or physically, staying in contact only in ways or at times when little or no emotional sacrifice is required of you.

By speaking honestly, you've done your part. Be wary of agreeing to anything that might rekindle the boundary-violation patterns of your childhood. Allow for the possibility of your feelings and your parents' changing in ways you could never have conceived of.

Telling people they hurt you can be an act of integrity. By speaking out, you give them an opportunity to hear specifically how they affected you. They may ignore your message, but at the very least, you've offered them a chance to take responsibility for their actions and make amends.

Stories of Confronting

Here are some ways in which the people I interviewed chose to confront their parents. In so doing, they balanced some of their childhood control.

A Turning Point: Ellen

Remember Ellen, the forty-nine-year-old volunteer worker whose Using, Abusing mom blamed her for the C-section scar from Ellen's birth? A turning point came when her mother began screaming at Ellen's five-year-old son because he moved Grandma's figurines. Ellen gathered up her three children and stormed out of her mother's home with the warning, "Don't ever scream at my children again."

Ellen had never before stood up to her mother so boldly. From then on, she felt empowered to demand civil behavior from her mother and to leave when she didn't get it. By protecting her children, she balanced a childhood in which no one was there for her.

First-Name Basis: Caitlin

Sometimes even a subtle change can have a profound impact. Remember Caitlin, the forty-one-year-old teacher whose Cultlike,

Perfectionistic mother wouldn't answer questions without the preface, 'Mom, may I speak?'" Recently, Caitlin began calling her mother, Patricia, by name.

"It was a conscious choice," Caitlin says. "It helped me see her more as a peer than a parent." Caitlin's commitment to telling the truth around her mother, no matter what the consequences, has balanced the speech control with which she was raised and has made honest communication possible.

Wrote to Her Father: Sharon

Thirty-one-year-old graduate student Sharon, in a letter, told her Smothering, Holocaust survivor father that he had failed to protect her against her stepmother's emotional abuse. "I wrote my letter over and over, read it to my friends, and kept revising it," she says. "I wanted to tell him I loved him but that I didn't like his behavior. I just wanted an apology." Her father wrote back, "How dare you question me! How could you do this to me? This is going to hurt my marriage."

"His tone was like God coming down from the sky," Sharon insists. Despite her father's reaction, she has stood her ground. In so doing, she's begun to balance years of emotional control: "Sometimes it's better to oppose and be angry, especially when a view of yourself has been imposed on you."

Confronting Abuse: Tess

Tess, a thirty-eight-year-old flight attendant, confronted her Depriving, Abusing mother on the physical and sexual abuse she suffered at the hands of relatives by placing a note on her mother's kitchen table reading: "Physical Abuse. Whippings. Molestation." When her mother saw the note "she just acted confused. She said that I had had a wonderful childhood. She painted some rosy picture of a middle-class upbringing. The abuse doesn't exist for her."

Though Tess's mother didn't acknowledge the extent of Tess's abuse, voicing her truth helped Tess feel she had honored her past and had defended herself: "It took me a long time to figure out that not everyone grew up like I did, with disapproval all the time, excessive discipline, and never being told they did anything well."

"Saw Fear in His Eyes": Will

Much of controlling parents' power is based on maintaining an exaggerated sense of size over their children. When adults visit bullying parents after a long absence, their parents often seem shorter, smaller, or less capable. The realization that their parents can no longer physically abuse them begins to reverse years of bullying.

Will, the twenty-eight-year-old teacher whose Perfectionistic father terrorized him for years, found freedom at age nineteen when he realized that his father could no longer beat him up. During a heated argument, Will raised his fist in his father's face. Although no blows were exchanged, Will recalls, "I felt ecstatic. It was freeing. He had absolute physical and economic power over me for so long. For the first time, I saw fear in his eyes."

"You Broke My Soul": Jorge

Sometimes a confrontation with a parent happens spontaneously. Jorge, the thirty-two-year-old psychiatric aide raised by an Abusing, Chaotic mother who whipped, kicked, and burned her children, turned a corner at age twenty-two when he saw his mother chasing his younger sister with a stick. He grabbed his mother's arm and said, "You just don't do that, that's inhuman." His mother struggled to get free and screamed at Jorge but he held tight. Jorge looked in his mother's eyes and said, "I know this tune, Mother, and you can scream insults as loud as you want, but nothing will take my hand away."

By standing up to his mother, Jorge achieved equal footing with his childhood tormentor. By preventing abuse of his sister, Jorge began to balance the lack of having anyone to intervene on his own behalf.

Coping with Anger: Samantha

Sometimes a confrontation isn't with one's actual parents but with one's anger over parental sins or omissions. Samantha, the forty-year-old artist who was physically abused by her Depriving mother and molested by a relative, testifies that much of her initial healing involved expressing anger: "A big step was in realizing how much rage I held in my body. It was making me depressed. I had a punching bag in my living room. I'd hit my tennis racket on pillows. I'd hit branches against trees or hit an aluminum garbage can with a baseball bat. I remember one day I took a two-by-four and hit it with a hammer until it was virtually toothpicks."

At first Samantha felt scared by her anger. "It felt bottomless," she admits. Then she went through a "they-did-this-to-me" stage. Later, with the help of her therapist, she realized, "My parents cost me a lot, but part of me still wanted to have some sort of relationship with them." In time, she found additional channels: writing, dancing, singing, painting, and sculpting.

Potential Risks of Confronting:

- Can be emotionally stressful
- May inflame relations with your parents
- May lead to retaliation from your parents
- May lead you to feel disappointment or grief, depending on the reaction you receive

Potential Benefits of Confronting:

- Can free you of the energy drain from carrying around unexpressed feelings
- Can clarify your options for relating to your parents
- May lead to positive changes in your relationship with your parents
- Can empower you by balancing a childhood of speech control

Exercises for Speaking up and Confronting

There are countless ways in which to speak out. Confrontation can be actual or symbolic. You may find that one or more of the following forms of symbolic confrontation can help you clarify what, if anything, you want or need to say to your parents.

1. **Write it down**. For those who had little freedom of speech growing up, writing your piece can heal by allowing you to freely express your grievances in a letter you do not send or in a conversation with a trusted friend or therapist. List what your parents did, what you wanted from them, how their control affected you—and how it may still affect you and those around you.

2. **Do a parental report card**. Remember how you had to show your school report cards to your parents for their signatures? Grade your parents on how well they:

   - Granted you emotional freedom
   - Saw you as a unique person and encouraged your potential

- Fostered open communication

- Modeled healthy boundaries and provided consistent limit setting

- Bestowed affection, acceptance, and physical touch

- Fostered an awareness of your inner life and encouraged connections outside the family

- Loved you, encouraged you to individuate, and set you free

3. **Have your day in court.** Visualize, write, draw, or role-play suing one or both of your parents. Imagine your lawyer cross-examining them about how they raised you, or cross-examine them yourself. You can do this exercise on your own or role-play it with a friend or therapist. The judge or jury (of your friends, if you like) then renders a verdict. In your closing statement, review the tactics your parents used and what they cost you. Demand an apology or any other compensation you choose.

4. **Have your day on *Oprah*.** Visualize, role-play, write, or practice with a friend or therapist being with your parents on *Oprah* or any other talk show. You are the focus of attention as you review your childhood. Your parents may be as controlling as ever, but you stick to your truth—and are applauded for it.

5. **Preach hellfire and brimstone.** You are the guest speaker at the First Church or Temple of the Controlling Parent. Show the congregation of controlling parents the error of their ways. Allow yourself to be as Bible- or Torah- thumping as you want. Visualize the congregation being attentive, afraid, persuaded, or whatever you choose.

6. **Stand-up comedy.** Envision doing a routine about your parents at a comedy club. Mimic their sayings and controlling behaviors as the audience roars and cheers.

19
CAN I FORGIVE
MY PARENTS?

People go too fast into forgiveness. Having enough time to feel angry was important to me. I had to blame. I had to feel like a victim.

—EVELYN, 46, A NURSE

I forgive my father because I understand him.

—SALLY, 31, A COMPUTER PROGRAMMER

Few issues cause more concern and confusion for those who grew up controlled than forgiveness.

I believe forgiveness is optional. Forgiving may aid healing or it may slow it down. For some, forgiving—and, more important, letting go—is freeing and healing. Others never forgive and still heal.

It takes courage to forgive, because it means letting go of part of your identity as a wounded person—a role that may have served to break denial and start your healing. Forgiving, then, may feel like you're abandoning hard-fought recognition of how you were wounded and what it cost you.

Yet it also takes courage not to forgive, if done consciously, in order to explore your feelings so that you can set them free. Doing this may mean that you have to tolerate many difficult feelings on your way to a resolution.

Many myths surround our conceptions of forgiveness.

Myth #1: Forgiving means forgetting.

Reality: You will probably always remember abusive parental control.

Forgiveness does not mean forgetting or excusing, nor does it mean denying your wounds. It means acknowledging a wrongdoing, experiencing the feelings connected with being wronged, and, after a period of time that only you can determine, letting go of actively holding the wrongdoing against the wrongdoer. Forgiveness includes letting go of a belief or illusion that things "should" or "could" have been different. Forgiveness can restore your general sense of trust and love to what it was before you were hurt, though you may never again fully trust the specific person who hurt you.

You may find it helpful to distinguish between the *content* of what parents said or did and the *intent* behind their actions, as Cocola and Matthews suggested in *How to Manage Your Mother*. Even though your parents may have hurt you, it's possible that their intent was to protect you, as an act of love.

Myth #2: Forgiving is the answer in any troubled relationship.

Reality: For some, forgiveness is unwise or impossible.

Forgiveness can be a trap, Forward writes in *Toxic Parents*. While it is important to let go of a desire for revenge, which can work against emotional well-being, you never have to forgive or absolve someone who betrayed you. Forgiveness often does not enhance healing and can even be a form of denial, writes Forward, who suggests forgiving only if the person who wrongs you does something to earn forgiveness, such as acknowledging what happened and seeking to make amends.

Myth #3: The sooner you forgive, the better.

Reality: Premature forgiveness can reinjure you.

Premature forgiveness can be especially injurious if it leads you to dishonor your feelings, ignore the truth, or do things for others that hurt your own best interests. These may be the very things you were forced to do in childhood.

Wayne Muller writes in *Legacy of the Heart: The Spiritual Advantages of a Painful Childhood*, "Forgiveness, while it may bring healing, has its own timing. It should be nurtured and invited, but never pushed. Any fear and rage must be honored and allowed to be true for as long as it is present. The heart knows when it is ready to forgive" (13).

Pressuring yourself to forgive can interfere with healing. It may be helpful to give yourself a grace period—six months, a year—with no pressure to forgive. During that time, you may attain forgiveness or you may not. But resolution comes more freely without pressure; you were pressured enough growing up. Parents or friends may become irritated with you for not being ready to forgive or for choosing not to forgive. "Let bygones be bygones," they urge. Their comments, however well-intentioned, often reflect their own discomfort rather than your needs. Never forget, your timetable is your own. Nobody else can determine it.

Myth #4: Forgiveness doesn't count unless you tell the person you've forgiven.

Reality: Forgiveness can be done silently or proclaimed verbally. What counts is that *you* hear it.

As a child, you may have been prevented from making choices that were in your best interests. Forgiveness is just such a choice. It may or may not include continued contact: You can cut contact with your parents and still forgive them; you can remain in contact and never forgive them.

You may want to forgive only after receiving a parental commitment that from now on your relationship will be respectful. You can hold your relationship with your parents to the same standard you hold other friendships; if it's a two-way relationship of trust, respect, communication, and acceptance, it's worthwhile. Otherwise, forget it.

Myth #5: Forgiveness is done for others.

Reality: Forgiveness is most freeing when it is done for you.

Your goal is to find greater peace and relationships that nurture you. Forgiving or not forgiving is an act of self-interest, not something you "should" do because it's "right." Sometimes, not forgiving can cause pain because it leads to suppressing your love for your parents, which Bloomfield in *Making Peace with Your Parents* calls a core need. "By holding on to . . . resentments, [we] surrender control over [our] own emotional well-being to the person who hurt [us] in the first place," Bloomfield writes (28).

Myth #6: Forgiving is a permanent act that takes away the hurt.

Reality: Forgiving is not all or nothing.

Forgiving doesn't mean you will never again feel turmoil about what was done to you. You may seesaw, feel sorry for your parents, realize their hardships and limitations, then remember the full extent of their mistreatment. It's important to take your time, explore your feelings, and protect yourself along the way, as therapist Mike Lew suggests in *Victims No Longer*. Forgiving is a pardon, not an exoneration, he writes, and it isn't all or nothing—**you can forgive a little.** Few people totally complete the task of forgiveness, even when they want to.

Forgiveness is a process with its own twists and timing. It's important to let the process unfold and have faith that it will do so. Muller's words can be a helpful guide:

What we are forgiving is not the act—not the violence or the neglect, the incest, the divorce or the abuse. We are forgiving the actors, the people who could not manage to honor and cherish their own children, their own spouse, or their own lives in a loving and gentle way. We are forgiving their suffering, their confusion, their unskillfulness, their desperation and their humanity (11).

Letting Go

Emotionally letting go can be something that is more helpful to focus on than forgiveness. Letting go means making relative peace with your feelings and memories of being hurt. Seen in this light, forgiveness is only an optional method for letting go. It helps some let go; it doesn't help others. You can let go by forgiving; you can let go without forgiving—it's your ball game. It's possible simply to overlook parental abuse and remain loyal. It's also possible to withdraw and blame. But both paths involve little conscious choice because they are *reactions.*

Bear in mind that it's of the utmost importance to honor yourself. You were forced into things as a child; don't force yourself into an artificial timetable now. A period of limited contact with controlling parents may or may not be wise. Some people can let go only after achieving a safe distance from parents; others can let go while living with their parents. Setting good boundaries between you and your parents, of course, helps the letting-go process. It's harder to forgive someone by whom you still feel engulfed or rejected. And as I've said, seeking a supportive sounding board is crucial to healthy separation.

There is no easy way to measure when you've mourned enough. Give yourself enough time to explore feelings so scary they went underground. Some people know viscerally that they are not ready to forgive, just as others know when they have been identifying for too long with a "victim stance" in a way that is more constricting than healing. It can be difficult to differentiate between the discomfort that comes from grappling with forgiveness and the discomfort of being emotionally stuck for too long. Trust yourself.

Stories of Forgiveness

Here are some ways in which those interviewed faced the issue of
forgiveness and letting go.

"Slow and Gentle": Evelyn

Evelyn's process of forgiving her Abusing father was "slow and
gentle." The forty-six-year-old nurse eventually forgave her father as
well as her Childlike mother who allowed the abuse, but she is glad
she allowed herself to take as long as she needed: "I feel strongly that
people go too fast into forgiveness. Having enough time to feel angry
was important to me. I had to blame. I had to feel like a victim."

Evelyn's anger on the subject of forgiveness became a useful
barometer: "As long as I still got angry when someone mentioned forgiving, I knew I wasn't ready to forgive." Eventually, anger faded and
acceptance came in.

By letting forgiveness happen at its own pace, Evelyn balanced her
father's bullying. By letting herself explore all aspects of forgiveness
before making a choice, she balanced her Childlike mother's emphasis
on always having to be certain.

Mother's Request for Forgiveness: Brenda

Her mother's request for forgiveness made the difference for
Brenda, the fifty-four-year-old homemaker who was the daughter of
grimly serious Perfectionistic parents who criticized her when she
laughed or was happy. Brenda's mother went into therapy and subsequently asked her daughter for forgiveness for emotionally abusing
her. "That was really important to me because I was willing to forgive
her if she was at least trying," Brenda says. "I believe we all do the best
we can. Given my mother's horrible childhood, she didn't have much
to do the best with."

Understanding Opened the Way: Sally

For Sally, the thirty-five-year-old computer programmer, understanding paved the way to forgiving her Smothering father who
refused to acknowledge her coming out as a lesbian: "I forgive my
father because I understand him. I value a lot of what I got from him.
I am a lot like him. I am subject to some of the same pitfalls and I try
to take a forgiving attitude toward myself." Over time, Sally has found
herself less reactive and more compassionate toward her father. "I
haven't hated my father in a long time," she says wistfully. "Mostly I

feel sad for him. He doesn't have the tools to be happy. He's been waiting to die since he was fifty."

Sally foresees eventually taking care of her parents if they become
too ill to care for themselves: "I can't imagine saying, 'Time to go to the
nursing home. See you at the funeral.' I'd like to think I'd find a balance between their needs and mine."

Will Not Forgive: Deirdre

Deirdre, the thirty-six-year-old office manager whose Perfectionistic, Cultlike stepmother brutally controlled her, will not forgive. "My
stepmom thinks she has a good relationship with me," Deirdre says.
"It's untrue. I've built a big wall around me and she doesn't get inside
anymore. I don't tell her anything I don't want her to know."

A key resource in Deirdre's healing has been conversations with
her sister: "She and I have two-hour conversations about how angry we
are with our stepmother. It's nice to have a sister as an ally."

"When my stepmom dies, I think I'll feel unchained," Deirdre confesses. "A few years ago, I probably couldn't have said I'd feel relief at
her death. But now I can say it and I don't have a sense of guilt.

"I try to tell myself she is who she is. She won't change, and the
past can't change. I can't confront her because she denies everything.
Maybe someday I'll forgive her. But I am still angry and I cannot forgive her now."

Struggles with Forgiveness: Patty

With both parents dead, Patty, the fifty-three-year-old counselor,
struggles with forgiveness: "I cannot forgive my father for humiliating
me sexually or threatening to burn me with a cigarette. I'd feel like a
traitor to that little girl who suffered all those things if I forgave him."

Instead of concentrating on forgiving her Abusing, Depriving
father, Patty focused on healing the pain he caused: "Forgiving my
father would honor him, which I am not ready to do. Focusing on healing myself honored me."

Not Ready to Forgive: Rosemary

Rosemary, the fifty-five-year-old manager whose Abusing, Using
mom would punch her, then rehearse her daughter in telling outsiders
that she'd walked into a door, freely states, "I am not ready to forgive.
It is heart-wrenching but I enjoy denying my parents me. Why should
I be there for them because they want me around? Why should I forgive and forget? I have wished my mother dead many times. That's the

saddest thing a child can say about a parent, but it's true. I don't miss her. I can't say I love her. I want to, and sometimes I do say 'I love you' and mean it, but my other feelings are so strong."

Internal Parents: Magda

Magda, the thirty-six-year-old civil servant whose immature, Using father got her birthday toys he wanted to play with, tries to focus on forgiving her internal rather than actual parents: "The parents in my head may not be the actual parents I had, but they're the archetypes. I may never be able to entirely forgive my actual parents but I can do anything I want with the parents in my head. They are the ones I *can* forgive."

Potential Risks of Forgiveness:

- If premature or forced, can emotionally reinjure you or slow your healing
- Can be a form of denial, rationalization, or minimizing
- Can lead to disappointment over lost illusions and unfulfilled expectations
- Cannot prevent future attacks or control by parents

Potential Benefits of Forgiveness:

- Can help you let go of hurts and emotionally move on
- Can lead to greater peace, energy, and freedom
- May open the way for a better relationship with your parents
- Can lead to greater self-acceptance

20

CAN I ACCEPT
MY PARENTS?

*Parents . . . are sometimes a bit of a disappointment to their children.
They don't fulfill the promise of their early years.*

—ANTHONY POWELL

Acceptance is another choice in relating to controlling parents. This doesn't mean accepting continued control, nor does it mean forgiving past parental hurts. But none of our relationships in life is perfect; all have limitations. Acceptance means seeing your parents accurately, honestly viewing the positives and negatives in your relationship, and choosing to continue the relationship.

Part of acceptance is understanding how different your parents' viewpoints may be from yours. Monica McGoldrick in *You Can Go Home Again* imagined how even Cinderella looked from her stepmother's perspective—"her Goody-Two-Shoes behavior could drive you to drink"—or from the perspective of Cinderella's stepsisters—who didn't match cultural dictates that women be "small, beautiful, gentle, long-suffering and unassertive" like Cinderella (279). You may not agree with a parent's perspective, but it is helpful to understand it.

As with confronting and forgiving, accepting tends to be most powerful when done primarily for yourself. For some people, acceptance can bring peace, since most of us want to honor, love, and esteem our parents. Casting parents out of your life may thus leave you feeling less than whole. For others, however, accepting or honoring an abusive person may feel like it repeats the abuse. Like confronting and forgiving, accepting is optional.

Events can facilitate acceptance: You become a parent; one of your parents dies or becomes ill; or there is a confrontation or rapprochement. Forging a better-defined sense of yourself can expedite a transformation. Yet events and time don't always make acceptance easier. Your relationship may get worse as your parents edge closer to death—the ultimate reminder of their lack of control. "Age is not renowned for improving the personalities of rigid, unhappy people," Steven and Sybil Wolin write in *The Resilient Self* (104).

Accepting often includes multiple recognitions: Your parents are flawed; you are flawed; your parents have hurt you; you have hurt your parents; you sometimes resent them or feel angry with them; and you sometimes feel love or closeness with them. Holding all these feelings and denying none can be a healing balancing act, especially for those from families in which conflicting feelings were not tolerated.

As I've said, no parent is all bad. Every mother or father, no matter how controlling, had moments of courage, sacrifice, responsibility, and love. Many controlling parents, despite the incredible damage they may have inflicted on you, also gave you a great deal. If nothing else, they gave you your life. It may be difficult to see your parents as both hurtful and caring rather than as seeing them as only hurtful, but a full-palette view of others tends to make *you* feel more whole as well.

It can be hard to accept that you cannot change your parents or their ways. If they are depriving or critical, you may be tempted to try and beat them at their own game. Yet trying to change them or beat them at their own game is a losing proposition for you since their game is one of winning, controlling, and not needing others. You can't change them, you won't beat them, and you probably won't get them to see the "error of their ways." But you can individuate, and this may open the way for greater acceptance of your parents just as they are.

Acceptance is based in reality, not on wishful thinking. Remember: Many controlling parents cannot maintain a stable sense of their own or their children's identities. Expecting a Using parent to be whole-heartedly generous without strings attached—like expecting a Smothering parent to respect your differences or expecting a Perfectionistic parent to have compassion when you are less than perfect—is like expecting a hungry grizzly bear not to eat you. Sometimes the bear will pass you by, just as your parents may sometimes pleasantly surprise you, but this is not the norm. Part of acceptance comes from having realistic expectations and protecting yourself accordingly. It's best to assume that bears will be bears.

Stories of Acceptance

Here's how some of those interviewed faced the issue of acceptance.

A New Understanding: Celina

Celina, the thirty-seven-year-old teacher whose schizophrenic
mother raised her in a terrifying, chaotic home, reached a new understanding in her late thirties about her mother, whom she had not seen
for years. Teaching, working in a psychiatric hospital, and doing magic
shows for children helped Celina heal from her nightmarish childhood. Especially helpful was a women's support group. Through the
women's group, Celina found "my heart began opening to my mother."

At about the same time, Celina gave birth to a daughter, and
became a single parent like her mother: "When I was pregnant, I realized how much my mother must have loved me. That was a turning
point. I was alone with my baby and I could see how hard it must have
been for my mother with two kids, no husband, and mood swings—
and they didn't talk about mental illness then."

When Celina finally located her mother on the New York streets
through homeless advocates, "She wouldn't even look at me at first.
She pushed me away. But the killer control she'd had was gone. She
had softened. She had lost that vicious, I'm-going-to-force-you manner."

Celina spent twenty-two hours with her mother, taking pictures,
making tape recordings, staying up all night talking. Her mother sent
her away with a bag of flyers on UFOs that she'd been collecting.
Recalls Celina, "I felt this great love and forgiveness. I tried to get her
to go to a shelter or senior housing but she refused. It was so heart-
breaking."

Celina created a dance piece from her visit with her mother: "I
found a respect for her lifestyle and choices. She was going through life
in a way that made sense to her." Sometimes when Celina thinks of her
mother, she dances. It brings her a measure of peace.

In Her Best Interests: Samantha

After years of estrangement, Samantha, the forty-year-old artist
whose Depriving, Abusing mother threatened to leave her behind in
stores for asking to go to the bathroom, decided that it was in her best
interests to resume contact. "I realized I'd never have a successful relationship with a man until I reestablished a relationship with my family, so for selfish reasons I began writing letters to them," she says.

In recent years Samantha has begun visiting her parents: "They
take me to dinner and we actually have fun sometimes. My mother is
no more nurturing than she ever was. She's still overcontrolling. But
she's not mean anymore. There is a definite wall there, but my parents

are trying. They say, 'I love you,' and I don't really care if they mean it or not. It's nice to hear it."

Part of Samantha's acceptance has come from letting go of her hopes of having nurturing parents: "There's always going to be a little girl in me who wants loving parents, but I know it's stupid to try and get nurturing from them. I would never call my mother and say, 'Mom, I don't feel well.' I'd be inviting her to abuse me once more."

For Samantha, forgoing her unfulfilled hopes in favor of an acceptable reality helped balance the pain of her past.

Healing in the Twilight: Margaret

Margaret, the thirty-three-year-old family-law attorney raised in the shadow of the Smothering, Perfectionistic father who rewrote her college admission essays and never let her win arguments, found acceptance in the twilight of her father's life. "In his later years my father was very apologetic for his excesses," she remembers. "Prior to that, I distanced myself from him and would not forgive him. In his final years I tried to understand him. He became dear to me."

Her father grew more interested in listening to Margaret: "We'd talk and talk. I stopped being a Republican, like he was. He didn't tolerate that well but at least he wanted to hear why. I really loved him and thought we were kindred souls."

By the time Margaret's father died, she had reached relative peace with him. It has taken longer, however, to let go of the legacy of his control: "If you have controlling parents, they are still controlling even after death. I still feel accountable to him."

Potential Risks of Acceptance:

• If premature or forced, can emotionally reinjure or disempower you

• Can be a form of denial or rationalizing

• May bring disappointment when you let go of your hopes and accept reality

Potential Benefits of Acceptance:

• Can lead to greater wholeness and peace

• Can create goodwill and a better relationship with your parents

- Can allow you to see different perspectives
- Can allow you to move on emotionally

Exercises for Acceptance

1. **Walk in their shoes.** Visualize your parents at two stages in their lives: when they were children and when they were young parents. Perhaps look at old photos of them. Seeing them as hurt children or scared young adults may offer you helpful perspectives.

2. **Reverse roles.** Imagine you are a controlling parent and your parents are now your children. Imagine controlling them. Notice what feelings you have. You may get a glimpse of both the power and pain they may have felt when controlling you.

21
SHOULD I REDUCE OR BREAK CONTACT WITH MY PARENTS?

My mother is not a part of my life anymore. I don't keep my distance out of meanness to her. I do it to protect myself and my children.

—ALICE, 42, A WRITER

For some adult children of controlling parents, the most viable choice is to completely break or radically reduce contact with one or both parents—either for a period of time or indefinitely—because some parents are so abusive that contact with them is "like walking into a propeller," wrote Victoria Secunda in *When You and Your Mother Can't Be Friends* (308).

Even when it's your healthiest choice, a complete break with a parent hurts. In order to take care of yourself, you are saying good-bye to someone who created you. For some who grew up controlled, it can be helpful to know that you can someday resume contact if things change. For others, keeping alive such a possibility leaves the door open to too much second-guessing or fear.

Stories of Reducing or Breaking Contact

Here's how some of the people I interviewed faced the question of reducing or breaking contact with a controlling parent.

Makes Her Life Miserable: Caitlin

Caitlin, the forty-one-year-old teacher whose navy officer father tyrannized his children with military-style discipline, has little contact with her father and anticipates less in the future. "He is in regular contact with my sister and makes her life miserable," she says. "He never takes an interest in my life." Caitlin reluctantly decided, after many demoralizing attempts to elicit her father's interest, that she would no longer try to maintain contact: "My predominant feeling about my father is sadness. I have missed him my whole life. He was such an intelligent and talented person, yet so screwed up. Sometimes I'll be watching a family-oriented TV show where they're so bonded and there's so much love and I get cynical. But really, I am feeling sad."

For Caitlin, balancing meant being willing to accept and live with her sadness instead of continuing to suffer by trying to reach her father.

No Contact: Carolyn

Carolyn, a thirty-five-year-old woodworker, no longer maintains contact with her Depriving, alcoholic father: "I'd always get off the phone feeling like a bad person just as I did as a kid. I keep thinking that somewhere inside my father is this loving, nurturing person who'll say, 'Honey, I know you had a hard life.' But that is like Charlie Brown and Lucy, who pulls the football away from him year after year.

"I don't even give him my phone number," she adds. "I can't give him the tiniest piece of information about myself without a critique coming back. I feel sorry for him but I feel less sorry when I think about that little girl he slammed around and called lazy and stupid."

Becoming More Distant: Alice

Alice, a forty-two-year-old writer, has increased both emotional and physical distance from her Using mother: "I send her cards and a Christmas gift, but I don't spend much time around her. I've never confronted her, never said, 'Change or I'll leave.' She'd never admit to doing anything wrong, so limited contact is the imperfect solution.

"I know I'll never get her approval," Alice admits. "I try to live without being hurt by her and realize it's okay that I don't like her. My mother is not a part of my life anymore. I don't keep my distance out of meanness to her. I do it to protect myself and my children. I have few regrets, though much sadness, about my choice. After she's gone, I'll probably be kinder in my memories."

For Alice, balancing has meant focusing on taking care of her own needs after a childhood of attending to her Using mother.

"Couldn't Live There": Colleen

Colleen, the thirty-three-year-old graduate student who was the oldest of seven in a Smothering family that insisted on utter conformity, left home for good at seventeen. Like many who grew up controlled, Colleen vividly recalls the moment she decided to leave.

"After Dad slammed me against a wall and broke my necklace, I couldn't live there anymore. I went to a friend's house whose mom co-signed on an apartment for me," she says. "I became dead to my father. He would never discuss me. He took me off insurance policies." Her only lifeline to her family came when her washing machine broke and her mother secretly took Colleen's laundry to their house, then sent it back with one of her sisters along with some lemon meringue pie.

In retrospect, the break made it easier for Colleen to individuate. "When a parent says you are dead and off the insurance policies, there's less of a hold on you," she advises.

For Colleen, balancing meant confronting her fears about eco-nomic survival.

"Jesus Loves You": Shirley

Remember Shirley, the forty-four-year-old artist whose fundamen-talist mother banned Christmas after realizing the word "Santa" had the same letters as "Satan"? Shirley fell into years of addictive drug use that started in her twenties, and lived on the streets until she got treat-ment. In hitting bottom, she realized there was a connection between her problems and her upbringing. After six years clean and sober, Shirley began setting limits on phone conversations with her mother by insisting she would hang up if her mother tried to proselytize: "When I told her no witnessing and no preaching, my mother lost it and screamed, 'You are just a junkie. God knows what is good for you. You don't know.'"

Shirley realized that the "good mommy" she wanted was never going to be there: "Every once in a while I feel really sorry for her. She is a three-year-old trapped in an adult body. I see her fear. I see how lost she is in life. But she was the mother, not me. I didn't bring a child into the world and punish it its whole life and expect good things to come of it."

Shirley feels she gave her mother a chance to make amends. Since her mother chose not to, Shirley feels finished with the relationship: "I feel at peace about not talking to her. It is a service to both of us."

Potential Risks of Breaking Contact:

- Can lead to an "emotional cutoff" that may leave you feeling less whole

- Can lead to retaliation from your parents
- Can bring feelings of guilt, disloyalty, grief, or loss

Potential Benefits of Breaking Contact:

- Can offer protection from further control
- Can provide a safe distance for healing
- May be the least costly among imperfect choices

Exercise for Saying Good-Bye to a Parent

Sit somewhere peaceful where you won't be disturbed. Envision your parent and yourself in a place where you feel safe, so that you can bring closure to the relationship. Unlike what happens in an actual encounter, your envisioned parent hears what you say and does not speak unless you want her or him to do so. Say everything you want to say so that you'll feel closure even if you never see your parent again. You might tell your parent how he or she hurt you and/or thank your parent for how she or he helped you. Fully confess your feelings. Then tell your parent good-bye.

If during this exercise you need time to compose your thoughts, imagine having your parent step outside. The point of this exercise is not to do or say it "right" or to worry about a parent's reactions. This encounter is for you. The exercise may allow you to find peace with both your actual and your internalized parents, who wield great influence even if you no longer have contact with your actual parents.

You can prepare for this exercise by writing in a journal or discussing with a trusted friend or therapist what you would want to say to your parents. During and after this exercise you may experience many emotions: grief, relief, anger, regret, resentment, freedom, and/or peace. Be compassionate with yourself about your reactions and acknowledge your courage in working to free yourself.

Next: Quandaries

Certain issues—parental aging, money, siblings, and holidays—can be especially challenging in balancing. The next chapter offers guidance on handling them.

22

FAMILY QUANDARIES

The family spirit has rendered man carnivorous.

—FRANCIS PICABIA

Certain issues can be especially challenging in balancing your relationship with controlling parents:

- Facing your parents' mortality
- Adult-life relationships with siblings
- Financial ties with parents
- Holidays and family rituals

Facing Parental Mortality

Faced with the awareness of parents' aging, many feel a pressure to get problems with their parents worked out before it is "too late." Such pressure can make setting new boundaries or reducing contact evoke feelings of disloyalty, even if it is the healthiest move.

It's important to avoid duplicating your role as a controlled child—that of satisfying your parents' needs before your own in order to avoid their wrath. If you want contact with a parent and can figure out a way to do it so that you gain more than you lose, your path is clear. If you don't want contact with a parent, or if you want contact but can see no way to be in contact without losing more than you gain, hold your ground. If you cannot yet resolve the dilemma, your best choice may simply be to have an awareness of the dilemma and proceed with your life until clearer options evolve.

You can only make choices based on what you know and feel now. The key is to do or say what you must, regardless of the response. It can be upsetting if a parent dies before you've worked it out or had a chance to say your piece. But, even after they're gone, you can still say what you have to say in a letter, meditation, or poem to them.

I've found people who grew up controlled worry that if there is going to be a rapprochement, it's entirely up to them. This may be a disempowering double standard. There is no "have to" or "should" about your relationship with your parents. Few people die with finished relationships. If you act in your own best interests and later come to feel you acted mistakenly, you may be sad. Yet if you're not acting in your own best interests, you're probably already sad. Have compassion for yourself. It won't help to add self-blame to what is one of the most difficult issues of your life.

A handful of those I interviewed said they expect to feel relief when a parent dies but expressed guilt for these feelings—a not uncommon dilemma. On the one hand, you may need time and distance from one or both of your parents or realize that reaching out would only invite further abuse. On the other hand, you may wish for more contact. Feeling unable to give up hopes of connection with parents but also feeling unwilling to return to a relationship where you will be hurt, you wait. In this situation, the prospect of a parent's eventual death can bring a sense of relief—along with grief, guilt, and loss—because it promises to end this state of limbo.

It was not until her parents died—her father seven years ago and her mother four years ago—that Patty, the fifty-three-year-old counselor, allowed herself to come to grips with her father's physical and verbal abuse: "When my father died, I felt relief. Yet he did love me as best he could. When he died, it was one less person on earth who loved me, and that was sad."

Exercise for Facing a Parent's Mortality

Visualize giving your parent's eulogy or writing his or her obituary. For those who fear parents dying with "unfinished emotional business," this exercise can crystallize your feelings. Say exactly what you feel about your parent and how she or he affected you. Tell your whole truth, good and bad, including what your parent did badly and what he or she did well. By writing or privately speaking what you'd like to say or wish you had said at your parent's funeral, you can better clarify what you might want to say to your parent while she or he is still alive—or find greater peace with your parent's memory if he or she is no longer living.

Financial Ties to Controlling Parents

Some controlling parents use money or gifts as a way to express approval and love. If you separate from your parents and their gifts stop, you will probably feel as if their approval and love has also stopped, and on some level it probably has.

You may have conflicting feelings about financial ties to your parents. One woman, for example, recounted her mixed feelings when she gets an occasional twenty or fifty dollars from her mother. She admits, "I want to send it back but I can use it."

Another woman said of her Using mother, "She controls a substantial amount of money and I don't want to be disinherited. I've earned it. I figure I'm making her happy. I'm pretending to be a dutiful daughter by having dinner with her once a month. Even though I know I'm not a dutiful daughter, she's happy with it."

At the other extreme are people who grew up controlled but get no financial help from their parents and expect none. Rather than strings-attached control, people with this experience may struggle with a sense of deprivation. For example, a woman whose parents lived lavishly but shared little of it with the children feels hurt each time her father visits. He expects her to pay for dinner even though he is well off and she's struggling financially: "His narcissism really hurts. It reminds me of all I didn't get growing up." In some ways, her father's detachment has made it easier for her to separate emotionally, though the pain of deprivation is still great.

Family financial ties can be supercharged with guilt, secrecy, and anxiety, and everyone's situation is unique. Some people who grew up controlled may desire financial support from their parents or expect an inheritance and don't want to jeopardize it. Others may feel phony by disguising their true feelings about their parents in order to gain financially. Hiding your true feelings from parents in order to retain an inheritance isn't "wrong" any more than is refusing to take anything from parents to avoid feeling compromised. The key is to make a choice based on your values and find the solution that, however imperfect, honors your needs and standards.

Relationships with Siblings as Adults

As you emotionally separate from your parents, you may find that relations with brothers or sisters can be healing or upsetting—or both. If a sibling also felt controlled, you may be able to compare notes and validate each other's experiences. If a sibling loyal to your parents gets mad at you for "making trouble" or tries to convince you to deny your reality, it can exacerbate your wounding.

Your siblings cannot emotionally separate from the family before they are ready, just as you could not. Brothers or sisters may have had a different experience than you did while growing up. They may not want to give up illusions about the family. They may be afraid of one or both parents. They may fear validating your position because they would feel unbearable guilt about not having protected you. If you were the child most targeted by parents, your siblings may feel guilty for receiving less abuse than you did.

If you have broken or reduced contact with a parent, it may be hard to listen when a sibling talks about a birthday party or holiday visit to Mom or Dad. So remember, it's normal to feel left out—even when you choose not to participate. If a sibling sees visiting a parent as a privilege, and you see it as a trauma, it's hard not to feel estranged.

You may be able to talk with a sibling about this. Ideally, you both can feel heard and "agree to disagree." But this may not always be possible. Part of emotionally leaving home may include emotionally separating from a sibling.

Sometimes connecting with siblings can aid healing in surprising ways. One woman whose mother pitted her against her brother in childhood had no contact with him for twenty years. When she reestablished contact, she found him even more controlling than her mother: "That set to rest any doubts that I grew up in a dysfunctional family."

By contrast, another woman was astounded when her younger sister told her how her emotional support had kept her alive years earlier when the younger sister had been a suicidal teenager. Until the revelation, the woman had never known the extent of her sister's pain or how supportive she herself had been. Since then, the two have become closer.

Holidays and Family Rituals

Holidays and family rituals are laden with emotional tugs. Parental birthdays and Mother's and Father's Days are supposed to honor parents, but how can you feel good about honoring someone who dishonored you? Year-end holidays may feel like a time for families to be festive, but how can you feel festive coming together in an environment of control, dishonesty, or manipulation?

Despite the roadblocks, holidays offer opportunities for you to observe your level of individuation and act in healthier ways. For example, deciding how much and what kind of contact you want with your parents at holidays—based on what is healthiest for you rather than on historical practice—can be an empowering step. One woman and her mother have agreed on a "Christmas truce" from Thanksgiving to mid-January, during which they don't talk about family issues or emotionally laden subjects.

It may help to *expect* that a visit to your parents will be stressful so you'll be less disappointed if you feel pressured. And if the visit isn't stressful, you're bound to be pleasantly surprised.

Exercise for Holiday Angst

Choosing cards for parental birthdays and holidays can be tough. It's hard to buy the ones that say "You were always there for me" if your parents weren't. Humorous cards might be misinterpreted. Buying a flowery, bland card may feel like selling out. You have so much you were never allowed to say, so why say something without any meaning?

One solution is to create and even send a more honest controlling-family card. Perhaps something like: "Even Though You Hurt Me, Thank You for What You Gave Me," or "Overall, It Was Still Better Than Being in an Orphanage."

Resources for Balancing

Bloomfield, Harold. *Making Peace with Your Parents.* New York: Ballantine Books, 1983.

Cocola, Nancy, and Arlene Matthews. *How to Manage Your Mother: Skills and Strategies to Improve Mother-Daughter Relationships.* New York: Simon & Schuster, 1992.

Cohen, Susan, and Edward Cohen. *Mothers Who Drive Their Daughters Crazy: Ten Types of "Impossible" Moms and How to Deal with Them.* Rocklin, CA: Prima, 1997.

Engel, Beverly. *Divorcing a Parent.* New York: Ballantine Books, 1990.

Engel, Lewis, and Tom Ferguson. *Hidden Guilt: How to Stop Punishing Yourself and Enjoy the Happiness You Deserve.* New York: Pocket Books, 1990.

Farmer, Steven. *Adult Children of Abusive Parents.* New York: Ballantine Books, 1989.

Secunda, Victoria. *When You and Your Mother Can't Be Friends.* New York: Dell Publishing, 1990.

Step Three: Redefining Your Life

23

NINE POWERFUL PATHS FOR GROWTH AND HEALING

Be not afraid of going slowly; be only afraid of standing still.

—CHINESE PROVERB

Emotionally leaving home and balancing your relationship with your parents can help lay the groundwork for Step Three: *Redefining your life* in terms of who you are and where you want to go, not in terms of your parents or your past.

Here are nine paths to growth and healing that I have found particularly valuable for those who grew up controlled. They focus directly on undoing the distortions of power, size, feeling, thinking, relating, and identity that come with controlling territory. For each path, I offer exercises that my clients and I have found helpful, along with suggestions for further reading. You might test one or two paths and see if they benefit you. And, please, don't fall into the perfectionistic trap of thinking that you must do all nine or that you have to do them perfectly!

The nine paths:

1. Identify and pursue your passions.

2. Make a place for yourself in the world.

3. Use your feelings as allies.

4. Deepen connections with others without losing your sense of self.

5. Identify and change thought patterns that limit you.

6. Pursue greater self-acceptance.

7. Live in the present.

8. Seek peace with your body.

9. Reduce your need to control life and others.

1. Identify and Pursue Your Passions

Whatever you can do, or dream you can, begin it. Boldness has genius, power and magic in it.

—GOETHE

We are fluid, not static, beings. At any given moment each of us is either growing, maintaining, or shrinking in terms of our sense of self and personal power. One helpful measuring tool is a simple self-assessment you can use anytime, anywhere, particularly when you are feeling confused, self-blaming, or under assault. Ask yourself, "Right now, am I growing or shrinking?" Notice what makes you grow. Growth generally comes from facing challenges; feeling seen and heard; giving to others in a balanced way; meeting or exceeding your expectations; being creative; and, perhaps more than anything, pursuing what you are passionate about. Notice what makes you shrink. Noticing can help you identify and alter constricting behavior or situations and open the door to growth-oriented behaviors.

Exercises

1. **Unlock your psychological clamps.** Spaceship launching pads have powerful clamps that hold rockets down for a few moments even after ignition until a critical launch force is built up. If these clamps ever malfunctioned and didn't let go, the spaceship would blow up or burn up. Perhaps many of your hesitancies and fears about why you can't or shouldn't follow your dreams are nothing more than psychological safety devices that have been holding on for too strongly or for too long. List the fears, beliefs, and expectations that keep you from exploring your dreams or passions. Then ask yourself, "If all of my fears disappeared, what would I pursue?"

2. **Learn from your heroes.** Pick out a real or fictional character whom you've always admired or envied. What part of the character's life intrigues you? What aspects of them do you want to emulate?

3. **See the deeper purpose of your actions.** Each of us can benefit from discovering a purpose larger than personal concerns to motivate us in the face of life challenges. In fact, you can discover a greater purpose from even your smallest tasks. Once a day, at work or elsewhere, pause and ask yourself: "What am I creating?" Answer at least seven times, each time responding with a larger definition of what you are creating.

    For example, when going to work: 1) I am creating a fresh workday; 2) My work creates rewards for myself and others; 3) Rewards create energy for health, happiness, intimacy, and play; 4) Health, happiness, intimacy, and play create more fulfillment; 5) Living as more fulfilled creates the impetus to nurture others; 6) Nurturing others creates success and expansion in others' lives; 7) Success and expansion in others' lives benefits the entire community both now and in the future

    And so on.

4. **Take a risk.** List some intimidating adventures you have thought about but never done, such as skydiving, bungee jumping, river rafting, horseback riding, taking a helicopter trip, or scuba diving. Pick one, find a safe, reputable company, and take the risk.

5. **Live your dreams.** List twenty-five things, big and little, that you have always wanted to accomplish or experience. Prioritize them and begin to execute them.

6. **Envision ideal days in your life five, ten, and twenty-five years hence.** Notice the key activities and experiences that would make those days ideal. Then draw up a plan for making these key elements part of your everyday life in some form.

Resources

Anderson, Nancy. *Work with Passion: How to Do What You Love for a Living.* San Rafael, CA: New World, 1995.

Bolles, Richard Nelson. *What Color Is Your Parachute?* Berkeley, CA: Ten Speed Press, 1997.

Sher, Barbara, and Annie Gottlieb. *Wishcraft: How to Get What You Really Want.* New York: Ballantine Books, 1986.

2. Make a Place for Yourself in the World

Every society honors its live conformists and its dead troublemakers.

—MIGNON MCLAUGHLIN

If you grew up controlled, you grew up without a Bill of Rights. Now's the time to assert your rights in the world.

If you felt small and powerless around your parents, you may feel small and powerless around others in your work or personal life—especially authority figures or controlling people. Exercises one through five below can help you develop more assertiveness.

Making a place for yourself in the world also includes giving yourself permission to express yourself in new and different ways. Exercises six and seven give you ideas on how to do this.

Exercises

1. **Develop your arsenal of self-defense.** If another person abusively criticizes you, you have several options:

   a. Confront it. Say, "That sounded like an attack. Was it meant to be?" Another approach is to ask, "Would you repeat that?" After they do, say, "That's what I thought you said," and then say no more. Your lack of reaction can knock critical people off their stride and silence them.

   b. Play dumb, asking questions endlessly so attackers have to repeat and clarify their criticisms. By putting the explanatory focus on them, you increase your own power and dilute their attacks.

   c. Respond with a non sequitur. For example, if someone critically asks why you did something, give a completely nonrelated answer such as, "My dog is due for her rabies shot." Or distract by completely changing the subject. Or ask an attacker a question you know she or he wants to answer. Or simply agree with them and move on.

2. **See bullies or critics as caricatures.** Visualize a person who unfairly criticizes you as a bantam rooster strutting about, a brat throwing a tantrum, or a scared rabbit huddling in the corner. Visualize responding. With each word you say, the critic becomes

smaller and fainter, as if you were turning a light-switch dimmer, until he or she vanishes. These techniques can balance power and size distortions you may have held since childhood.

3. **Make at least one mistake a day.** Pick something that isn't dangerous or crucial and intentionally flub it. Pass a great parking spot and find one farther from your destination. Park crooked. Mispronounce a word in a meeting. Say "I don't know." By deliberately making mistakes, we see that the consequences of failing are generally benign or far less dire than we fear. Realizing this affords greater freedom to risk and persevere.

4. **Be a controller too.** If your parent was a bull in a china shop when it came to recognizing nuances in life, temporarily adopt that role in an exaggerated way. Find a cooperative person or situation where nothing is at stake and practice being obstinate, bullheaded, and simplistic. Notice if it gives you more energy than worrying and second-guessing yourself. Notice, too, if it drains energy from those around you. Ask others how it feels to be around such a controller. Let their responses validate how stressful it was for you as a child.

5. **You're entitled.** When your rights are violated, imagine that the violation is happening to your best friend or an innocent child. If you could intervene for them, wouldn't you? If so, give yourself the same benefit of the doubt.

6. **Try new forms of self-protection.** Try martial arts, tai chi, self-defense training, or assertiveness training.

7. **Try new forms of self-expression.** Pursue classes or hobbies in the expressive or creative arts, such as singing, dancing, painting, pottery, acting, public speaking, writing, or poetry.

Resources

Butler, Pamela. *Self-Assertion for Women.* San Francisco: HarperSF, 1992.

Evans, Patricia. *The Verbally Abusive Relationship: How to Recognize It and How to Respond.* Holbrook, MA: Bob Evans, 1992.

Napier, Nancy. *Getting Through the Day: Strategies for Adults Hurt as Children.* New York: W. W. Norton, 1993.

Smith, Manuel. *When I Say No, I Feel Guilty.* New York: Bantam Books, 1985.

3. Use Your Feelings as Allies

Sometimes a scream is better than a thesis.

—RALPH WALDO EMERSON

Since many controlled children had their physical needs better taken care of than their emotional ones, feelings can be one of the most important areas in which to seek balance. Your parents may have lopped off some of your emotional limbs. By letting your feelings branch out, you'll grow stronger.

Emotions give you physical clues as to their identity. You may feel a pulsing in your ears as you get angry; a tightness across your jaw or chest when you are afraid; quickened or slowed breathing when you are worried. Use these "hints" to alert you to oncoming feelings so that you can attend to them rather than shutting them off.

It's also important to honor your sensitivity, especially if it was squashed or ridiculed by a controlling family. Controllers tend to be uncomfortable with others' sensitivity and send messages that sensitivity is a flaw or a sin. Yet sensitivity to feelings, to others, and to yourself is truly a gift. Accepting and promoting your sensitivity can be healing after a lifetime of being shamed for it.

Exercises

1. **Take the lid off.** Since controlling parents often forced you to bottle up your anger, sadness, grief, rage, and joy, you may find it helpful to take off your emotional lid. Even if it seems forced or "pretend" at first, find a safe setting and scream, sing loudly, hit pillows, bang garbage cans. (Be sure to warm up your voice or body prior to this so you minimize the chances of strain or injury.) Taking the lid off extends your emotional range.

2. **Name that tune.** Recall a recent situation in which you felt strongly, such as a reaction to a movie, TV show, or book. Pause for a few moments and notice as many subtle variations in your feelings and sensations as you can identify. Take your time. Don't worry if it seems like several feelings are bound together. Observe each of those feelings, sitting with each one for a full minute.

3. **Expand your emotional range.** Avenues include acting classes, rebirthing and holistic breath work, various types of body work, journal or poetry writing, painting or drawing.

4. **See what new environments can evoke in you.** Watch children or animals playing, or a good movie. Volunteer to help feed the homeless on a major holiday. Notice the range of emotions you see in those around you. Take stock of what that evokes in you.

Resources

Aron, Elaine. *The Highly Sensitive Person.* New York: Broadway Books, 1996.

Goleman, Daniel. *Emotional Intelligence.* New York: Bantam Books, 1997.

Lee, John, and Bill Stott. *Facing the Fire: Experiencing and Expressing Anger Appropriately.* New York: Bantam Books, 1995.

Lerner, Harriet. *The Dance of Anger.* New York: HarperPerennial, 1985.

Rubin, Theodore. *The Angry Book.* New York: Collier, 1993.

4. Deepen Connections with Others Without Losing Your Sense of Self

It is easier to live through someone else than to become complete yourself.

—BETTY FRIEDAN

Intimate relationships can be difficult if you grew up controlled. Just as your parents' control strategies came one relationship too late—they controlled in response to how they were raised, not in response to you—your strategies to avoid control may not be helpful with your friends and loved ones.

Growth lies through balance: keeping a healthy sense of self, reality-testing your fears and perceptions, and allowing yourself to recognize that your childhood was then and this is now. If you grew up in social isolation, even a single, corrective relationship based on trust and respect—whether it is with a mate, friend, therapist, or coworker—will help you make great strides in undoing a lifetime of control.

Exercises one through five offer ways to maintain your sense of self in relation to others. Exercises six through nine offer ways to reach out and touch others.

Exercises

1. **Practice saying no.** Say no at least three times a day, particularly to offers, requests, or situations that don't benefit you. There are many ways in which to say no, ranging from polite but firm ("Thank you but I'd rather not"; "I'm sorry, I cannot"; "I'm afraid not"; "No thank you") to emphatic and resolute ("No"; "I'm not interested"; "That doesn't work for me"). Each time you say no, you put in a boundary that was violated when you were a child.

   Then practice saying yes. Say it at least once a day in situations that are safe but in which you might normally say no—particularly about things you've wanted but haven't felt entitled to.

2. **Not just smokers can step outside.** If you feel yourself becoming lost, small, or depressed in the company of others, take a break. Stepping outside, taking a brief walk, or finding someplace to sit quietly and tune into yourself helps restore a balanced sense of self.

3. **Be "disagreeable" from time to time.** In conversations with others, allow yourself to uphold an opposing opinion if it reflects your beliefs. Don't allow yourself to grow quiet or withdrawn simply because you are the only one voicing a certain view.

4. **Break the cycle.** Family therapist Virginia Satir suggests visualizing three families: your parents' families when your parents were children; your family when you were a child; and your own children when they were (or may someday be) born. What's different among the three families? What's the same? What do you want to make sure does and does not get passed on? What do you hope your descendants will say about you?

5. **Tally your boundary setting.** Notice each time you do not allow abusive or controlling people to abuse or control you. Each victory validates your progress. Even returning a defective purchase to a store for a full refund can be an instructive exercise.

6. **Acknowledge our interdependence.** We carry within us all the teachers and friends who taught us and loved us; the animals we have befriended; the plants and animals we eat, wear, and use; the thou-

sands of craftspeople, food growers, businesspeople, researchers, caregivers, entertainers, news gatherers, government employees, and so many others who make our lives easier and safer. While you may have had controlling parents who did you damage, you also carry within you the gifts of thousands of people who bear you no ill will and some of whom genuinely loved you. It may help to remind yourself of the interdependence we all carry as humans and visualize or thank those who have helped make your life easier and more fulfilled.

7. **Cultivate nurturing surrogate family members.** It is normal to hunger for connections to older people and earlier generations; they give us a context for our lives. If you can't connect with your own parents, you may be able to connect with relatives. If not, you can volunteer at a senior center or "adopt" a friend's parents or a neighborhood elderly person. These elders can be the parents or grandparents you wanted but without the controlling baggage. In turn, you'll be providing the elderly with company and nurturing. Other avenues include volunteer work at hospices, hospitals, Big Brothers or Big Sisters, or community centers.

8. **Touch.** If you have difficulty expressing physical affection, start slowly and build on it. Hold or hug objects, then animals, then perhaps children, then yourself, then others. Next, practice receiving touch from each of these (Farmer, 141). Yes, even trees can hug back, even if only in your imagination. Try this a few minutes a day for a week.

9. **Experiment with giving and receiving.** Allow yourself to ask for something from a trusted person. Let yourself experience the out-of-control feelings that can accompany wanting, asking, and receiving. Likewise, do a favor for someone and note any controlling and out-of-control feelings the act produces in you. Balancing giving and receiving helps avoid your always being the one in control.

Resources

Bradshaw, John. *Healing the Shame That Binds You.* Deerfield Beach, FL: Health Communications, 1989.

Whitfield, Charles. *Boundaries and Relationships.* Deerfield Beach, FL: Health Communications, 1993.

5. Identify and Change Thought Patterns That Limit You

He who masters the grey everyday is a hero.

—Fyodor Dostoyevsky

Attribution theory is a potent technique for recognizing and changing inner distortions—and, in so doing, redefining your life. It comes to us from cognitive therapists such as Aaron Beck, David Burns, and Martin Seligman, who suggest that how we think of ourselves, others, and events determines how we behave and feel.

We tend to view events in one of two ways:

1. **All-inclusive** (affecting our entire lives), **permanent** (forever unchangeable), and **innate** (caused by inborn personal traits).
   or
2. **Limited** (confined to a specific circumstance), **temporary** (a passing event or changeable situation), and **external** (caused by situations outside ourselves).

For example, one person on a diet who eats a pint of Ben & Jerry's Chocolate Fudge Brownie ice cream thinks, "It's hopeless. I'll never lose weight. I have no willpower." Yet, another dieter thinks, "I had one slip. It's a setback, but my diet is still in effect. Things must be getting to me today."

The difference between these two dieters is in how they explain their behavior. Dieter Number One attributes all-inclusive *(It's hopeless)*, permanent *(I'll never lose weight)*, and innate *(I have no willpower)* reasons for eating the ice cream. Dieter Number Two attributes limited *(I had one slip)*, temporary *(It's a setback, but my diet is still in effect)*, and external *(Things must be getting to me today)* reasons for the same act. The first explanation is likely to leave the dieter feeling pessimistic, depressed, and guilty, thereby causing him or her to lose the motivation to continue dieting. The second explanation leaves that dieter feeling sobered but without guilt, depression, or pessimism, and therefore more likely to continue or even redouble positive efforts.

It's highly likely you grew up awash in unhealthy attributions. Controlling parents tend to see their children's troubling behaviors as **all-inclusive** ("You never do anything right"), **permanent** ("You'll never amount to anything"), and **innate** ("You're lazy and stupid"). Yet when their children do something well they offer **limited** ("Doing that right still doesn't prove you're not stupid"), **temporary** ("I doubt you can do that again"), and **external** ("You couldn't have done it without my help") reasons.

Ironically, controlling parents tend to see their own behavior as just the reverse: attributing their own successes to innate, permanent, and all-inclusive factors, while explaining their failures as temporary, limited, and caused by external reasons—such as their children.

Healing lies in reality-testing your attributions. If your parents tended to externalize the source of their parental problems by saying they were your fault, you grew up internalizing negative beliefs about yourself. While these negative internal beliefs may feel real, they are usually untrue.

When faced with a problem or troubling feeling, ask yourself: Is this a limited or an all-inclusive event? Temporary or permanent? External or innate? *Externalizing* the source of your *problems* (as limited, temporary, and external) and *internalizing* the source of your *successes* (as all-inclusive, permanent, and innate) can be a prescription for healthier attitudes. Of course, internalizing all positives and externalizing all negatives can be nothing more than wishful thinking if practiced blindly and without reality testing. Yet if you grew up controlled, it's likely that you have been uncritically internalizing negatives and externalizing positives. If you've fed yourself an endless diet of self-blame, some wishful thinking may be helpful. Eventually, you'll find a healthier balance in realistic thinking.

Exercises

1. **Give back family myths.** Your parents never really knew you because they couldn't see you as separate, judged you by their standards, and tagged you with their fears. Be on the lookout for myths and lies you hold about yourself. List the ways in which you have differed with and broken from controlling-family habits. Write down the ways in which you are fundamentally different from each of your parents.

2. **Neutralize brainwashing.** Recall uncomfortable situations, such as high-pressure insurance or service-contract pitches, car or electronics buying, or being panhandled. Observe the elements of coercion used. Be cognizant of any untruths, threats, plays upon guilt, and misuses of power. Remember how you felt. This helps you identify parallels in your family and in your inner dialogue.

3. **So what?** Fears are often mental pictures of an imagined end result. Thinking about jumping off a cliff might bring a rush of pleasure if we didn't have a picture of ourselves smashing into the rocks below. When you have a fear, ask yourself: "Okay, if the worst happens, then what?" Keep asking yourself this question until you get to the ultra worst-case scenario.

   This will allow you to 1) reality-test whether your feared end result is an actual, probable, or even likely occurrence; 2) contain overwhelming feelings by spelling out your fears so you can see that they are finite; and 3) take your fears to the ultimate conclusion and realize that you can trust yourself to find ways to deal with the event.

4. **Sharpen the noggin.** Take courses in or read about logic and communication. Do crossword puzzles or play games like chess, hearts, or spades. If you don't have friends who like to play these games, you can find plenty of free sites on which to play them with cyber buddies on the Internet. At any age, we can create an infinite number of new dendritic connections—those branchlike endings of brain neurons. The more the connections, the greater your mental abilities will grow.

Resources

Burns, David. *The Feeling Good Handbook*. New York: Plume, 1989.

Mallinger, Allan, and Jeannette DeWyze. *Too Perfect: When Being in Control Gets Out of Control*. New York: Ballantine Books, 1992.

McWilliams, Peter. *You Can't Afford the Luxury of a Negative Thought*. Los Angeles: Prelude, 1995.

Seligman, Martin. *Learned Optimism*. New York: Knopf, 1991.

6. Pursue Greater Self-Acceptance

Most of us have a surgeon's mentality when it comes to those [parts of our] selves we dislike.

—HAL AND SIDRA STONE

Since parental control distorts how we see ourselves, we come to feel at odds with aspects of ourselves that we are ashamed of or fear.

Yet shunting off those parts leaves us feeling fragmented. Every aspect of you has a contribution to make. You may not like some aspects of yourself, but accepting their presence leaves you stronger than denying them does.

One or more of the following exercises may help you reclaim aspects of yourself you have cast off.

Exercises

1. **Know the shadow.** List any people you strongly dislike, marking down the qualities about them that you dislike most. These very qualities may reflect, as Carl Jung suggested, your disowned or "shadow" parts. Ask yourself honestly whether any of these annoying qualities exist in you. Another approach is to think of things you've done that trouble or embarrass you. Objectionable others and embarrassing memories can become teachers, showing you parts of yourself with which you struggle. Realize that all parts of you, both the ones you like and the ones you don't, make you who you are.

2. **Have a "parts party."** Each of us carries within us subpersonalities, such as protector, perfectionist, nag, critic, hero, child, feminine, masculine, dreamer, comedian, rebel, bully, pessimist, black sheep, and genius. Satir advocates holding an imaginary "parts party" to which you invite all these subpersonalities. By mingling with your subpersonalities or role-playing their parts, you may learn and accept more about yourself.

3. **Keep a dream journal.** A foundation of Jungian psychology, the dream journal is a way to discover your less-than-conscious feelings and strivings. For a week, keep pen and paper by your bedside and, immediately on waking, write down any dreams you remember. Then, on your own or with others, write or talk about your dreams and what they might represent. Dreams are like art; much is expressed, and there are infinite valid interpretations.

4. **Try new avenues.** Various writing techniques can give voice to the quieter parts of you. One is freewriting, in which you write or type nonstop at a comfortable pace whatever comes into your mind for fifteen to twenty minutes. Anything you write is okay; don't worry about grammar, punctuation, or spelling. The only rule is to write nonstop whatever comes into your mind (even if you have to write

"I can't think of what to say" a dozen times until something else comes to mind). Afterward, in reading over what you have written, you'll see new connections and kernels of insight or creativity.

Another approach is to take pens in both right and left hands and write a dialogue. Envision the dialogue as being between your heart and mind, the two sides of an impending decision, or any two parts of yourself (Capacchione). After a time, astounding statements may emerge from your nondominant hand.

5. **Use a favorite social or political movement as a metaphor for your self-inquiry and healing.** Writer Deena Metzger used the example of "personal disarmament." If nuclear disarmament is something you feel strongly about, ask yourself if there is a part of you that terrorizes other parts of you, if there are parts that war with other parts, or if you have emotional "nuclear missiles" piled away (Bogue). Similarly, if you're an environmentalist, do you have any distorted thoughts or habits that are clear-cutting, strip-mining, or dumping toxic waste in your life and need halting or cleanup? Are there precious resources in your body, mind, or soul you are wasting?

6. **Convene your inner board of trustees.** Imagine a meeting of your inner board of trustees with you as chairperson. Envision calling to attendance your various roles and parts: career self, emotional self, body, mind, sense of humor, self as citizen, mother/father, son/daughter, brother/sister, wife/husband, yourself as a child, your future self, and/or any other aspects of yourself you like. Set the agenda—perhaps goals for increased happiness and nurturance. Then ask each "member" of the board for input on your life and goals.

7. **Open yourself to a new group activity.** Much of the benefit from women's groups or from the men's movement comes from embracing a part of ourselves that we possess but do not fully know or appreciate. We can often become better at self-acceptance by seeing how others have accomplished greater self-acceptance. Groups can also help us acknowledge who we are to others.

Resources

Branden, Nathaniel. *The Six Pillars of Self-Esteem*. New York: Bantam Books, 1994.

Carson, Richard. *Taming Your Gremlin.* New York: HarperPerennial, 1983.

McKay, Matthew, and Patrick Fanning. *Self-Esteem,* 2nd ed. Oakland, CA: New Harbinger, 1992.

Stone, Hal, and Sidra Stone. *Embracing Our Selves: The Voice Dialogue Manual.* Novato, CA: Nataraj, 1989.

7. Live in the Present

I only ask to be free. The butterflies are free.

—CHARLES DICKENS

Life only happens in the present, and what-ifs and if-onlys are not present-based. If you focus on what-if (the future), you tend to be anxious. If you focus on if-only (the past), you may live with regret. Living in the present gives you the greatest opportunity to be a full participant in your life, living each moment as it occurs. Grounding yourself in the here and now is richly rewarding, especially if you lived much of your childhood dissociating.

Great novels, great poetry, and great film all deal in specifics. They may address huge themes but do so with specifics: colors, textures, nuances, small actions. So, note specifics in your life. Speak in specifics. Think in specifics.

Observe pets. Dogs and cats are remarkable for their ability to live in the present. When a dog is happy, he or she shows it completely. When he or she is afraid, it shows. When dogs and cats are hungry, they eat; when tired, they sleep. The moment an appealing activity appears, dogs and cats seem to completely forgive, forget, and move on. We can learn from animals.

Exercises

1. **Stop, look, and listen.** Take twenty minutes and remove all distractions. Turn off the phone or turn down the answering machine. Absorb yourself in your thoughts, feelings, bodily sensations, and desires. Observe what is in your immediate environment: clothes, furniture, walls, floors, trees, water. After a time, you may notice how much calmer you feel—and how much you're focused in the present.

2. **Meditate.** There are many forms of and purposes for meditation—relaxation, awareness, concentration—but sitting quietly for fifteen to twenty minutes a day does wonders. Envision rhythmic, gentle images such as warm white light, falling water, bubbles rising, circles on water, or snow falling. See your thoughts or worries as fleeting bubbles rising away from you. Repeat a simple mantra or tune into your breathing. Whichever method you choose, you're learning to be in the present and developing moments of inner peace.

3. **Take a "conscious" bath or shower.** Find a time when you are alone and not rushed and take a luxurious bath or shower. Tune in to each of your senses. Close your eyes and listen to the water dripping and cascading. What do the water, tub, soap, and your skin smell like? What do the tub and tile feel like against your skin? Look at the hues of color in the water, your skin, the shower. Gently and slowly touch various parts of your body, reveling in the various sensations from your hand and the part of the body that is being touched.

4. **Sit with nature.** Sit outdoors and be sensitive to textures, smells, sounds, scenes. Allow this practice to expand gradually to your everyday life so that you'll fully appreciate even the simplest of activities.

Resources

Carlson, Richard. *Don't Sweat the Small Stuff . . . and It's All Small Stuff.* New York: Hyperion, 1997.

Hanh, Thich Nhat. *The Miracle of Mindfulness.* Boston: Beacon Press, 1997.

Kabat-Zinn, Jon. *Wherever You Go, There You Are: Mindfulness Meditation in Everyday Life.* New York: Hyperion, 1995.

8. Seek Peace with Your Body

Compassion for myself is the most powerful healer of them all.

—THEODORE ISAAC RUBIN

Your body is your home, the one thing you truly "own" in life. For those who grew up controlled, it's a challenge to treat your body in ways other than your parents did.

We have explored ways in which to ignore or reduce the critical thoughts you take in from your inner critic. By the same token, increase your attention to what you physically put into your body. Increase your education about nutrition. Every one of your cells can be nourished or diminished by what you eat. A healthy diet can reduce mood swings, depression, fatigue, irritability, and cloudy thinking. Get regular physical checkups and dental care and congratulate yourself on doing so.

Exercises

1. **A walk a day.** Take a daily walk, ideally for at least twenty minutes, though even five minutes can do wonders. Swing those arms. Walking is enlivening. It feels good to move our bodies through space and walking is one of the most natural ways to do it.

2. **Tune in, turn on, but don't drop out.** Stretching, progressive muscle relaxation, biofeedback, massage, bodywork, meditation, or yoga can help. Even a ten-minute nap can be restorative. The extra attention you give to your body pays off.

3. **An hour a day.** Former L.A. Dodgers manager Tommy Lasorda has said that out of the twenty-four hours a day, each of us deserves to give our bodies one hour. Exercise, stretch, do fitness training, aerobics, or yoga. If an hour seems like too much, start with twenty minutes. When you think of all the hours you spend not taking care of your body, an hour a day isn't so long.

4. **If appropriate, seek professional help.** If you are depressed, you may have a distorted body image and may loathe and/or punish your body. For some, a trial course of medication may help. Most of us readily take medicine to right a chemical imbalance in our muscles and stomachs or other organs when needed. Yet for many there is still a stigma attached to taking medications like antidepressants or antianxiety agents to right an imbalance in brain chemistry. We know that long-term stress or traumatic events can alter our brain chemistry, thus altering our moods. The stress and trauma of years in a controlling family can contribute to an imbalance in neurotransmitters, which can induce chronic, low-level anxiety or depression. For some, medication designed to restore a healthier chemical balance in our brains can make a significant and lasting difference. Taking medication is a personal decision, but it's worth trying if other avenues haven't brought the progress or healing you desire. If

you choose to try a course of medication, it's generally best to seek out a qualified psychiatrist or very knowledgeable physician.

A few things to keep in mind: All medications have side effects, though most people can find at least one medication that offers far more benefits than side effects; problematic side effects from antidepressants tend to recede with time; and it may take several weeks of working with a physician and trying one or more medications before you can tell if the medication will help you. Many people taking an antidepressant or antianxiety agent report that they feel **more** like themselves, not less like themselves or in an altered state. It helps to realize that events not of your own choosing may have altered your brain chemistry; one way to restore balance is a medication of your own choosing with a physician's supervision.

Resources

Anderson, Bob. *Stretching.* Bolinas, CA: Shelter, 1980.

Kramer, Peter. *Listening to Prozac.* New York: Viking, 1993.

Robertson, Joel, and Tom Monte. *Natural Prozac: Learning to Release Your Body's Own Anti-Depressants.* San Francisco: Hampers, 1997.

Rodin, Judith. *Body Traps: Breaking the Binds That Keep You from Feeling Good About Your Body.* New York: Quill/Morrow, 1992.

Weil, Andrew. *8 Weeks to Optimum Health.* New York: Knopf, 1997.

9. Reduce Your Need to Control Life and Others

The only person you can control is yourself.

—MARIAN WRIGHT EDELMAN

Control is the opposite of trust. People who grow up with the trauma of control often develop pessimism about the future and try to control life in order to avoid disappointment. While this was a survival tactic in childhood, it robs you as an adult of optimism about a joyful and challenging future.

Building trust can come from expressing gratitude, from prayer, and from noticing your strengths and successes, not just your failures or challenges. Every living thing has a spirit, self, presence, essence, higher self, innate consciousness—however you choose to define it. Search for and cultivate your transformative sense of *spirit*uality. It may come through religion or it may not. It may come through gardening or bike riding or volunteer work or exercise or loving others. You may have been raised with a certain religion or view of God or spirituality that doesn't fit what's in your heart and your spirit. Spirituality is about faith and trust. Control is about fear and mistrust.

Trust can take time to build, but each time you succeed or survive a challenge, you trust your abilities more. It can also help to distinguish between things you can change and things you cannot change. The "Serenity Prayer" from Alcoholics Anonymous offers guidance:

> God grant me the serenity
> To accept the things I cannot change,
> The courage to change the things I can,
> And the wisdom to know the difference.

Exercises

1. **Meet your future self.** When faced with a challenge or decision, envision yourself in five or ten years, then ask your future self for advice. Doing so underlines the faith you have in your own innate development (Napier).

2. **Trust yourself.** Go through an hour assuming that you are completely trustworthy, your feelings reliable, and your intuition accurate. As situations come up, ask yourself, "If I knew I was absolutely trustworthy, what would I do now?" This can help you see that you have within yourself all you need to handle challenges (Muller).

3. **Trust gravity.** One helpful exercise is Napier's "Gravity-is-your-friend" (70). Lie down and feel the support of the bed or floor. Feel all your weight ease down into it and gradually let the ease deepen for five minutes. The earth will support your weight, and gravity will keep you grounded. Trust it. You can take this experience of trusting into relationships and situations.

4. **Express gratitude.** Take a minute at the end of the day to recollect all the experiences and gifts for which you are grateful (Muller).

5. **Notice what you do.** For one week, each night before bed spend

five minutes listing what you accomplished, experienced, or became aware of that day. At the end of the week, look over your lists. You'll see plenty to acknowledge. This builds the inner nurturer instead of fueling the inner tyrants.

6. **Explore various paths to spirituality.** Pray. Meditate. Read. Visit a cemetery. Read about or visit Jerusalem or other "holy" sites. Explore existential philosophy. Attend various church services such as an inner-city gospel, a fundamentalist tent revival, a Catholic mass, or a New Age or Zen center. Go on a vision quest.

Resources

Frankl, Viktor. *Man's Search for Meaning*. New York: Touchstone, 1984.

Levine, Stephen. *A Year to Live: How to Live This Year As If It Were Your Last*. New York: Bell Tower, 1997.

Levine, Stephen. *Who Dies?* New York: Doubleday, 1982.

Moore, Thomas. *Care of the Soul*. New York: HarperCollins, 1992.

Muller, Wayne. *Legacy of the Heart: The Spiritual Advantages of a Painful Childhood*. New York: Simon & Schuster, 1992.

Wolin, Steven, and Sybil Wolin. *The Resilient Self: How Survivors of Troubled Families Rise Above Adversity*. New York: Villard Books, 1993.

24
MAKING MEANING

Leaves turn color so we will get the message that before we let go of the tree of life, we need to show our beauty.

—BERNIE SIEGEL, M.D.

Overcontrol is a lack of healthy love. By now, you've learned that your parents probably didn't get healthy love early in their lives and it distorted their worlds. Unable to give you healthy love, they distorted your world. Now it's your turn to break the cycle. To the extent that you can accept and love yourself, you can accept and love those around you. To the extent that you can accept and love those around you, you'll have less need to control them. By breaking the cycle of unhealthy control, you contribute not just to your mental health but also to the health of those around you as well as all those who will come after you.

As this book ends, we are left with certain paradoxes. You can grow, but you may always have some limits as a result of growing up controlled. You can heal, but you may continue to feel occasional hurt or sadness from your past. While you can achieve a large measure of freedom from growing up controlled, you may never completely stop hurting or feel totally free. Whether you confront, accept, forgive, set healthier boundaries, or break contact with your parents, the relationship may always hurt. Healing from growing up controlled can be a lifelong process.

Being free after growing up controlled doesn't mean you won't be affected by your childhood or your parents. You cannot erase what happened. Accidents happen. Wounds happen. If you cut your hand, ignoring it or pretending it doesn't hurt can be dangerous and will only slow your healing. Freedom comes from acknowledging your emotional wounds, understanding how they happened, observing how they hurt and limit you, and seeking healing.

There are no guarantees, no way to control life. Your parents tried that and failed. Instead of trying to empower themselves in the face of life's risks, they lived reactively, desperately trying to reduce the risks.

You don't have to do that. You can build up your strength and flexibility and relationships and self-love so that you are better prepared to face life's risks. While your life may have lacked happy beginnings, there is plenty you can do to increase the chances of happy endings.

One way is by *making meaning* of your past. Making meaning includes synthesizing both the helpful and hurtful from your upbringing. Making meaning isn't a matter of simply saying, "It was all for the best," nor is it a matter of disowning the past. Rather, making meaning involves acknowledging the pain in your past along with your strengths. After all, you showed strength even while you were being wounded. Think of it! You derived strengths from painful events. These strengths are just as much a part of you as are your wounds.

You survived a controlling upbringing because you found ways in which to cope, despite the lack of help and the enormity of parental control. You survived because you developed strengths borne of an unfair, painful situation. It may be hard to switch gears from focusing on wounds to feeling pride in your strengths. Yet both are truly part of you.

Though you had little power in a controlling family, you were more than a passive recipient of parental control. It's helpful to see yourself as both victim and survivor, as innocent wounded child and courageous, resourceful warrior, because you were all of them.

Never forget that in childhood you showed great resourcefulness. For example:

- If at some point you had the horrifying thought that something was wrong in your family, you showed courage and insight in recognizing it. Completely brainwashed people don't know they've been brainwashed.

- If you grew up feeling unloved, it hurt. But it also may have helped you emotionally leave home more quickly and completely than if you had had more moments of love.

- If you spent a lot of time alone as a child, you may have felt lonely or the development of your social skills may have suffered. Yet if your family was destructive and abusive, your choice to spend time alone showed innate wisdom. Better to be alone than to be abused.

- If you did things to please your parents even though you didn't want to, such as earning better grades or pursuing extracurricular activities, you still reaped the benefits of your accomplishments.

- If you got lost in your world of play, hobbies, reading, writing, or drawing—though these may have been the only ways you could escape parental control—your activities brought you pleasure and developed your creativity. You carry all those gifts with you today.

- If you've been hurt in relationships as an adult partly because of the unhealthy patterns your parents modeled, it's unfortunate. But it also shows that you have the courage to keep trying, to risk hurt even after a painful childhood.

- If you grew up full of turmoil, it may have fatigued, hurt, or depressed you. But it also shows that you were emotionally responsive—that your true self, highest self, innate self, or however you conceive of it was struggling with all that was heaped upon you. You fought internally the battles you couldn't hope to win in the external world of your parents. You showed you had an inner spirit that would not just roll over and submit. If you had done everything your parents wanted and accepted everything they said, you would have had little turmoil—yet you would have grown up with a dwarfed will and a diminished sense of self and individuality.

You were a young warrior in protection of your soul. That doesn't minimize your pain or how unfair it was. But it shows who you were and who you still are.

While Part Three of this book has outlined a three-step healing process of separating, balancing, and redefining, in actuality, healing from growing up controlled is anything but a step-by-step process. It is holistic. Over time, a newer self emerges. You will notice, perhaps several times a month, how your values are changing, your emotional range is increasing, or your thinking is becoming more flexible. It takes time to ease out of old lessons and habits and grow into new ones. But it happens.

Much of the initial work of healing lies in making your wounds real: remembering, uncovering, and feeling the hurt. During this initial phase, much of your identity may be that of a wounded person and your world view may be dominated by the realizations, memories, and feelings that came with that wounding. Yet at some point unique to each person, healing includes moving on. If your identity persists endlessly as that of a wounded family member, you become attached to, not free from, the wounding family.

You may have gotten stuck with the bill for several generations' problems, but ensure that the buck stops with you. Use the feelings forged in the trauma of growing up controlled. You grew up "brainwashed," with little access to information, few allies, and a tightly controlled regimen—and you survived. Knowing this can help you face just about anything.

THE BOOK IN A NUTSHELL

- Children *never* cause or are to blame for abusive control.

- Most unhealthy control is not deliberate or conscious. It reflected your parents' fears, upbringings, and limitations. It did not reflect anything about you.

- If your parents grew up with trauma and never got help, they may have learned to mistrust the world. Mistrusting the world, they may have felt they had to control everything around them, especially you.

- Your parents had one or more controlling styles (Smothering, Depriving, Perfectionistic, Cultlike, Chaotic, Using, Abusing, and Childlike). They also used many or all of the Dirty Dozen (control of food, bodies, boundaries, social life, decisions, speech, feelings, and thoughts, and bullying, depriving, confusing, and manipulating). Recognizing and identifying these styles and methods can help you gain distance from the lasting effects of parental control as well as cope with the critical messages from your internalized parents.

- Unhealthy parental control may have forced you to distort aspects of yourself. These distortions can be at the root of many adult-life problems. Though it can be hard work, uncovering and balancing distortions of size, feeling, thinking, relating, and identity can allow you to solve present-day problems at the source.

- Whatever themes of control you grew up with tend to rear their heads when you emotionally leave home. Notice when your inner tyrants' messages parallel the commands, beliefs, and styles of your parents. These messages tend to come with the territory of emotional separation.

- Compliment yourself on any efforts you make to individuate. If you grew up controlled, even the desire to emotionally leave home is a big step.

- Cultivate contradiction. Situations are rarely either-or.

- Our relationships with parents change over time. Even if you once had either a closer or a more distant relationship with a parent, months or years later new issues may surface that can bring new closeness or distance.

- Healing continues twenty-four hours a day. Every minute, once you have chosen to emotionally separate and balance, you are gaining distance and power.

- One of the opportunities in healing is struggling with the same issues that defeated your parents—and succeeding.

- Recognize that you are the head of your family. Your family may include close friends, relatives, a significant other, children, even pets. You have the opportunity to create a different family atmosphere than the one your parents created.

- Healing from overcontrol is a long-term process that can stir complex feelings and spark difficult choices. Have confidence that you will make the right choices. Many people have overcome the legacy of a controlling upbringing. There are many paths to healing.

- You don't have to heal alone—even if you were an emotional orphan for years. Seek support where you can.

- You cannot change the past, but you can influence your future. Even a small effort to reduce the effects of growing up controlled pays big dividends for you, those around you, and future generations.

- Be patient with yourself. The costs of control took years to create.

Suggestions for Parents Who Were Raised in Controlling Families

No matter how you were raised, you are not destined to repeat your parents' hurtful child-raising patterns. According to family therapist Michele Weiner-Davis, more than two thirds of adults abused as children do *not* abuse their own children.

Keep in mind:

1. There's no shame in being ignorant about healthy parenting skills. There is, however, tragedy in being unwilling to learn and get help as a parent.

2. It wasn't your parents' mistakes that did you the greatest damage, it was their unwillingness to face their mistakes and learn from them.

3. Your model for parenting may be distorted. Ask yourself which would have been healthier for you as a child and which would be healthier for your children:

   • Having a flawless parent or having a parent who sincerely apologizes for mistakes?

   • Having a brilliant parent or having a parent who praises children's efforts at brilliance?

   • Having a parent who never hurts children or having a parent who sometimes accidentally hurts children but consistently lets them know they are loved?

4. By attending to your *own* needs for growth and nurturing, you give your children a great gift. Pain from the lack of inner growth and nurturing can lead parents to overcontrol.

5. Despite your best intentions, you may sometimes do the controlling things to your children that your parents did to you. Have compassion for yourself.

Suggestions for Partners and Friends of Adults Raised in Controlling Families

1. You have the right to block abusive behavior that partners or friends who grew up controlled direct at you. You can reject their

behavior while letting them know you are not rejecting them as people.

2. Your partners or friends may want you to agree with their viewpoints about their parents, but what they want even more is to feel validated. You don't have to agree with their viewpoints, but you can validate them by letting them know you see their pain and are trying to understand their dilemmas.

3. If your friends or partners grew up controlled, minimizing or discounting their feelings may reopen their childhood wounds. Since people who grew up controlled rarely felt heard or seen, simply listening to them can bring tremendous healing.

4. Healing has its own timetable. Pressuring others to make choices or move on before they are ready to tends to be counterproductive.

5. Respect their privacy. You can offer to listen to friends and partners if and when they want you to, but leave it up to them to tell you about their feelings and their process.

6. Realize that although their upbringing may pose challenges in your relationship, it can also bring special gifts to the relationship.

7. Remember that trusting others can be especially difficult for those who grew up controlled. It is a sign of trust when friends or partners include you in their healing process by talking with you about it.

8. Try not to take personally their projections onto you (i.e., if they accuse you of being "just like" their controlling parent). Their background, not your behavior, is often the cause of their projections.

9. Be cautious about verbalizing things from their parents' point of view unless asked to do so. Otherwise, to them it may feel like a betrayal.

10. Recognize that healing is uncharted territory for them as well as for your relationship. Perfectionism and judgment aren't helpful. Healing has its phases. Trust them and yourself.

Bill of Rights for Those Who Grew up Controlled (And Everyone Else)

We hold these truths to be self-evident: All people have the right to:

1. Ask questions

2. Dissent

3. Confront, prevent, or remove themselves from others' abuse and unhealthy control

4. Feel all their feelings and express them appropriately

5. Develop their own values, thoughts, and goals

6. Learn, grow, and connect with others

7. Make mistakes, experiment, and be uncertain

8. Choose whom they associate with

9. Pursue happiness, success, and health

10. Love and be loved, trust and earn others' trust

11. Self-respect and to earn others' respect

12. Pursue their spirituality

13. Be here

Notes on Research

To recruit interview participants I posted notices at universities, libraries, other public gathering places, and computer on-line services asking for volunteers. I was stunned by the response. Calls came within six hours of the first posting and continued for more than twenty-four months. Twice as many people responded as I was able to interview.

In the interviews, we explored participants' upbringings; their parents' and grandparents' histories; the legacies of their parents' control; their current struggles to relate to their parents; and, most of all, how they've tried to heal. Participants also completed a lengthy follow-up questionnaire about a year after their initial interview.

This was not an easy process for those who volunteered. Several scheduled appointments only to cancel after having second thoughts. Others participated knowing that it might be uncomfortable. As one thirty-eight-year-old woman asked, "How much Kleenex should I bring?"

Several people brought family pictures or artifacts. One woman even brought a flowchart mapping out the mixed messages and guilt-inducing statements made by her mother that stymied her early steps toward independence.

Ultimately, most people seemed relieved by being able to talk. One fifty-three-year-old woman said after a four-hour interview, "I know I've talked nonstop but I was never, ever allowed to say anything growing up."

While the group of forty participants is not intended to represent the greater population in terms of cultural makeup or socioeconomic status, by many measures it was a diverse group. The forty adults interviewed ranged in age from twenty-three to fifty-eight. The average age was thirty-eight. Two thirds of participants were female, one third male. Half were either married or in stable relationships. A third were parents.

Three quarters were college graduates. More than half were working in professional, managerial, educational, or artistic fields. About half were in psychotherapy when interviewed.

A third of those interviewed were raised as Protestants, slightly more than a quarter as Catholics, slightly less than a quarter as Jewish, and the rest with little religious affiliation.

All but two participants were living in the San Francisco Bay area when interviewed, though most participants had grown up outside California. Nearly a quarter of those interviewed were either born outside the United States or had lived a significant part of their early lives abroad.

While this was primarily a group of white, middle-class professionals of Northern European descent, one out of five participants was from a minority ethnic or sexual culture. Two Latino/Latinas, one African American, one Asian American, and at least four gay men and lesbians were among the forty adults who participated. Of course, the role of parents and the meaning of "control" vary tremendously among African American, Asian American, Latino/Hispanic, and other cultural groups. While I was thankful for the breadth and richness of experience contributed by those who volunteered who were Latino or non-white, I make few generalizations about controlling parents in specific racial and ethnic groups. Participants' stories and insights should be considered anecdotal and not necessarily representative of their cultural group as a whole.

Sources for Statistics

1. An estimated one in thirteen adults in the United States has grown up with unhealthy control. There are at least two ways to extrapolate a reliable estimate of the number of controlling parents and their children:

Method #1. Based on reported cases of child abuse

More than 3 million cases of child abuse occurred in 1997, according to the National Committee to Prevent Child Abuse and the National Center on Child Abuse and Neglect. Of these, excess control has historically been shown to be a key factor in one third of child-abuse cases (Gil). That's 1 million cases of child abuse annually in which excess control is a key factor.

Taking into account historical figures of child-abuse prevalence and population size, and adjusting to avoid double-counting subsequent abuse of the same children, this suggests that nearly 13 million of the 199.5 million adults alive today were abused with excess control as children. This is 6.5 percent of the 1998 adult population.

Method #2. Based on estimated prevalence of mental disorders that can lead to a controlling style

Studies of the prevalence of mental disorders show that between 5.1 and 6.6 percent of the adult population has a mental disorder that would likely lead to controlling behavior (U.S. Census figures; the National Institutes of Mental Health 1992 Epidemiologic Catchment Area study; and studies by Swartz et al., Nestadt et al. (1991), Nestadt et al. (1994), Sanderson et al., Mavissakalian et al., Samuels et al., Resnick et al., Brom et al., Davidson et al., Breslau et al., Bourgeois et al., Bourdon et al., Oldham and Skodol, and Loewenstein). The applicable *DSM-IV* mental disorders include obsessive-compulsive disorder, posttraumatic stress disorder, social phobia, panic disorder, and generalized anxiety disorder; some mood and dissociative disorders; and the narcissistic, borderline, histrionic, schizoid, and obsessive-compulsive personality disorders.

Applying this figure to parents born prior to 1962 (the cut-off date for the vast majority of parents of those eighteen and older in 1998), adjusting to avoid counting mental disorders that occur among parents before or after their child-raising years, adjusting to avoid double-counting parents who have more than one disorder, and factoring in historic population and family size census figures, an estimated 13.9

million to 20.9 million adults alive today grew up with *at least one* controlling parent. This is 7 to 10.5 percent of the current U.S. adult population.

Using either child-abuse figures or adult mental-disorder figures, at least 6.5 percent (13 million adults) and as many as 10.5 percent (20.9 million adults) of the current adult population grew up with abusive control. A conservative estimate would be one third the difference between high and low estimates, closer to the lowest estimate. The result: 7.8 percent, or more than 15 million adults, grew up with at least one controlling parent.

I believe this figure is, if anything, on the low side. Much unhealthy control doesn't meet the legal definitions of "child abuse," and many parents who do not have a mental disorder nonetheless overcontrol their children. This estimate, however, provides at least a starting point for a discussion on unhealthy control of children.

2. One in twenty children has a parent die during their childhood. From Jeffrey Dolgan, Ph.D., chief of psychology, Children's Hospital, Denver. "The impact of loss on children," unpublished article, August 1995.

3. One in six children has a parent who abuses alcohol or is alcoholic during the child-raising years. Derived from the National Institute on Alcohol Abuse and Alcoholism "National Longitudinal Alcoholic Epidemiologic Survey," 1992, and the National Association for Children of Alcoholics "Facts About Children of Alcoholics," November 1995.

4. One in seven children has a parent who suffers serious depression or mental illness during the child-raising years. Derived from the National Institute of Mental Health "Epidemiologic Catchment Area Survey," 1992.

5. One in five children is physically or sexually abused. Derived from the National Center on Child Abuse and Neglect "Third National Incidence Study of Child Abuse and Neglect," 1996, and McCauley, Jeanne, et al., "Clinical Characteristics of Women with a History of Childhood Abuse," *Journal of the American Medical Association,* May 7, 1997.

WORKS CITED

American Psychiatric Association. *Diagnostic and Statistical Manual of Mental Disorders*, 4th ed. Washington, DC: American Psychiatric Association, 1994.

Bass, Ellen, and Laura Davis. *The Courage to Heal*. New York: Harper-Collins, 1988.

Bloomfield, Harold. *Making Peace with Your Parents*. New York: Ballantine Books, 1983.

Bogue, Nila. "The Choice of a Lifetime." *Noetic Sciences Review*, Summer (1994), p. 31.

Bourdon, Karen, et al. "Estimating the prevalence of mental disorders in U.S. adults from the Epidemiologic Catchment Area Program Studies." *Public Health Reports*, Vol. 107 (1992), p. 663.

Bourgeois, James, et al. "An examination of narcissistic personality traits as seen in a military population." *Military Medicine*, Vol. 158, 3 (Mar. 1993), pp. 170–174.

Bowen, Murray. *Family Therapy in Clinical Practice*. New York: Jason Aronson, 1978.

Breslau, Naomi, et al. "Traumatic events and posttraumatic stress disorder in the urban population of young adults." *Archives of General Psychiatry*, Vol. 48, 3 (Mar. 1991), pp. 216–222.

Brom, Danny, et al. "The prevalence of posttraumatic psychopathology in the general and clinical population." *Israeli Journal of Psychiatry and Related Sciences*, Vol. 28, 4 (1991), pp. 53–63.

Capacchione, Lucia. *The Power of Your Other Hand*. North Hollywood, CA: Newcastle Publishing, 1988.

Cocola, Nancy, and Arlene Matthews. *How to Manage Your Mother*. New York: Simon & Schuster, 1992.

Davidson, Jonathan, et al. "Posttraumatic stress disorder in the community: An epidemiological study." *Psychological Medicine*, Vol. 21, 3 (Aug. 1991), pp. 713–721.

Farmer, Steven. *Adult Children of Abusive Parents*. New York: Ballantine Books, 1989.

Forward, Susan, with Craig Buck. *Toxic Parents*. New York: Bantam, 1989.

Gil, David. *Violence Against Children*. Cambridge, MA: Harvard University Press, 1970.

Golomb, Elan. *Trapped in the Mirror: Adult Children of Narcissists in Their Struggle for Self*. New York: William Morrow, 1992.

Lew, Mike. *Victims No Longer*. New York: HarperCollins, 1988.

Loewenstein, Richard. "Diagnosis, epidemiology, clinical course, treatment, and cost effectiveness for treatment for dissociative disorders and MPD." *Progress in the Dissociative Disorders*, Vol. 7, 1 (Mar. 1994), pp. 3–11.

Mavissakalian, Matig, et al. "Correlates of DSM-III personality disorder in obsessive compulsive disorder." *Comprehensive Psychiatry*, Vol. 31, 6 (Nov./Dec., 1990), pp. 481–489.

McGoldrick, Monica. *You Can Go Home Again*. New York: W. W. Norton, 1995.

Middleton-Moz, Jane. *Children of Trauma*. Deerfield Beach, FL: Health Communications, 1989.

Miller, Alice. *The Drama of the Gifted Child*, rev. ed. New York: Basic Books, 1994.

.———*For Your Own Good*, 3rd ed. New York: Noonday Press, 1990.

Muller, Wayne. *Legacy of the Heart: The Spiritual Advantages of a Painful Childhood*. New York: Simon & Schuster, 1992.

Napier, Nancy. *Getting Through the Day: Strategies for Adults Hurt as Children*. New York: W. W. Norton, 1993.

National Committee to Prevent Child Abuse. *Current Trends in Child Abuse Reporting and Fatalities: The Results of the 1997 Annual Fifty State Survey*. Chicago, IL: 1998.

Navarro, Mireya. "Holocaust Survivors' Emphasis Is on Life." *The New York Times*, national edition, February 1995, p. 12.

Nestadt, Gerald, et al. "Compulsive Personality Disorder: An epidemiological survey." *Psychological Medicine*, Vol. 21, 2 (May 1991), pp. 461–471.

————"Obsessions and compulsions in the community." *Acta Psychiatrica Scandinavia*, Vol. 89, 4 (Apr. 1994), pp. 219–224.

Oldham, John, and Andrew Skodol. "Personality disorders in the public sector." *Hospital and Community Psychiatry*, Vol. 42, 5 (May 1991), pp. 481–487.

Resnick, Heidi, et al. "Prevalence of civilian trauma and posttraumatic stress disorder in a representative national sample of women." *Journal of Consulting and Clinical Psychology*, Vol. 61, 6 (Dec. 1993), pp. 984–991.

Samuels, Jack, et al. "Personality Disorders in the Community." *American Journal of Psychiatry*, Vol. 151, 7 (June 1994), pp. 1055–1062.

Sanderson, William, et al. "Prevalence of Personality Disorders among patients with anxiety disorders." *Psychiatry Research*, Vol. 51, 2 (Feb. 1994), pp. 167–174.

Satir, Virginia. *Conjoint Family Therapy*, 3rd ed. Palo Alto, CA: Science and Behavior Books, 1983.

Secunda, Victoria. *When You and Your Mother Can't Be Friends*. New York: Dell Publishing, 1990.

————*Your Many Faces*. Berkeley, CA: Celestial Arts, 1978.

Swartz, Marvin, et al. "Estimating the prevalence of borderline personality disorder in the community." *Journal of Personality Disorder*, Vol. 4, 3 (Fall 1990), pp. 257–272.

U.S. Bureau of the Census. *Census of the U.S. and Current Population Reports*. Washington, DC: 1990.

U.S. Bureau of the Census.. *Resident Population of the United States: Estimates by Age and Sex*. Washington, DC: 1998.

U.S. Department of Health and Human Services National Center on Child Abuse and Neglect. *Child Maltreatment 1994: Reports from the States to the National Committee on Child Abuse and Neglect*. Washington, DC: 1994.

U.S. Department of Health and Human Services National Institute of Mental Health. *Epidemiologic Catchment Area Survey*. Washington, DC: 1992.

Weiner-Davis, Michele. "Common Questions About Relationships." *USA Today*, Nov. 29, 1994, pp. 4D.

Wertsch, Mary Edwards. *Military Brats: Legacies of Childhood Inside the Fortress*. New York: Fawcett Columbine, 1991.

Whitfield, Charles. *Boundaries and Relationships: Knowing, Protecting and Enjoying the Self*. Deerfield Beach, FL: Health Communications, 1993.

Wolin, Steven, and Sybil Wolin. *The Resilient Self: How Survivors of Troubled Families Rise Above Adversity*. New York: Villard Books, 1993.

About the Author

DAN NEUHARTH, PH.D., is a licensed marriage and family therapist with a Ph.D. in clinical psychology. A popular speaker, college educator, and award-winning journalist, he specializes in helping adults cope with the challenges of unhealthy family control. He lives in the San Francisco Bay Area.

If you would like to contact Dan Neuharth, Ph.D.:

drdan@controllingparents.com

Visit the author's official website at *www.controllingparents.com*